Earth Repair Gardening

THE LAZY GARDENER'S GUIDE TO SAVING THE EARTH

Kate Wall

Photography and Editing, Heidi Caddies
Scribbles, Kate Wall

Earth Repair Gardening

Copyright © 2021 by Kate Wall.

All rights reserved. No part of this publication may be reproduced, distributed or transmitted in any form or by any means, including photocopying, recording, or other electronic or mechanical methods, without the prior written permission of the publisher, except in the case of brief quotations embodied in critical reviews and certain other noncommercial uses permitted by copyright law. For permission requests, write to the publisher, addressed "Attention: Permissions Coordinator," at the address below.

Kate Wall

thegardenerswall@westnet.com.au

Brisbane, Queensland, Australia

www.katewall.com.au

Book Layout ©2017 BookDesignTemplates.com

Earth Repair Gardening/ Kate Wall. —1st ed.

ISBN 978-0-6487318-2-5

Contents

Introduction 1
Sustainable Gardening 7
    Sustainabillity is different for everyone 8
    How sustainable is your garden? 8
    Organic versus sustainable 14
Consumerism 11
Some Gardening Basics 13
    Understanding your climate 13
    Understanding your site 19
    Get to know your soil 21
The Kerkin Garden - A tale of many microclimates 23
Sustainable Soil Care 29
    Organic matter in soil 30
    The role of soil care in earth repair 37
    The role of mulch in soil care 38
    Choosing mulch 39
    Applying mulches 45
    Putting it into practice 47
Garden Products for the Soil 49

- Soil additives 49
- Fertiliser 50
- Plant tonics 55
- Soil conditioners 57
- Potting mix 65

Water 67
- Drought proofing 68
- Hydrophobic soils 69
- Water sources 71
- Water wise gardening 76
- Irrigation 79
- Too much water 80

Emma's Garden – Taming a challenging landscape 85

Landscaping Vs Gardening 89
- Hardscaping and hard surfaces 89
- Landscaping materials 91

Herbicides and Weeds 103

Sustainable Lawns and Hedges 109
- Fertilising lawns 110
- Mowing 111
- Turf variety 111
- Lawn pests 112
- Lawn weeds 113

Power tools 114

Hedges 116

Chemicals – Pesticides and Poisons 117

Bugs and bees and birds 117

Fungal pests 124

Plant Choices 127

Plants growing wild 130

Climate appropriate plants 131

Sourcing plants 135

Growing Food and Medicine 139

Easy food 139

Soil contamination 143

The medicinal garden 144

Native Plants 149

Tania's Garden – Wild at heart 154

Pots 159

Plastic pots 160

Decorative pots 162

Repurposing in the Garden 165

Reconstructing old pallets 166

Old china 167

Chairs 168

Insect hotels 169

Pots 170

Baskets 170

Plant choices 171

Theme 171

Paint 172

Old tools 173

Children's gardens 173

Tyres 174

Trellises 175

Features 175

Using the Garden to Reduce the Eco Footprint of Your Home 177

Trees in the garden 180

Sarah's Garden – More than just a garden, it's a way of life. 184

Becoming a Garden Ecologist 189

Emulating nature 189

If in doubt, add compost.

—Kate Wall

Earth Repair Gardening

Introduction

We are often told that our eco footprint is too big and that we're not doing enough to reduce it. Most of us are very much more aware of the environmental challenges facing our world than ever before, and we are trying to find those small ways to do our bit. But in truth, we are still rather attached to our modern lifestyles. What if doing our bit was to create a garden that we loved? What if by tending our own small patch of earth, and growing our own little piece of paradise, we could do something real that positively contributed to the healing of our planet? Isn't it wonderful to think that saving Earth could be as simple and as beautiful as a garden? This book is all about teaching you to garden in such a way that your garden can be a significant part of your own carbon offset system. It will teach you to be a great gardener (if you're not one already!) and will show you how to do real and actual environmental good with your patch.

Much research is being done into the role of the soil in sequestering carbon from our atmosphere. There is real potential to repair the earth's damaged soils, and in doing so to reverse climate change. What an exciting idea! And what is the critical key to repairing soil? PLANTS!!!! Gardeners rejoice, we can save the world!

I have been telling people for many years that my job is to heal the planet, one garden at a time. As we grow plants, they draw carbon from the atmosphere and store it in the soil. The more plants we grow, the more carbon gets stored. We do not have to grow forests (although forests are needed too), we just need to grow plants, lots of wonderful green plants. The key word though is 'grow'. As plants die, they are no longer so good at storing carbon in the soil and depending on what we do with the dead plant material, can even release that carbon back into the atmosphere.

This book then is going to guide you through becoming an earth repair gardener. Whether you are a green thumb or a decidedly black thumb, you will see your plot of soil differently and you will want to get growing. In particular, the focus of this book is to teach you how to garden in harmony with nature. You will be gardening sustainably in a way that is easy and filled with success.

Why write this book?

I began writing this book a number of years ago. As I picked it up to finish it, I wondered if it still needed to be written. Without question we are so much more aware of the need to reduce our personal carbon footprints and climate change has moved into overdrive.

The world of gardening has changed too. Organic gardening is readily accepted and even expected. While garden shows may still attract sponsorship from large companies supplying inorganic fertilisers, it is hard to find a garden expert willing to talk anything but organic gardening. This was not the case 10 years ago.

Previously there was a handful of large chemical companies making all the fertilisers, pesticides and other garden treatments we could need. Now there is a huge proliferation of small innovative companies making all sorts of fabulous organic alternatives – from aloe vera liquid plant tonic to beneficial insects via mail order.

The plight of bees around the world has shone a spotlight on the decline of insects and the incredibly important role insects play in a garden. Gardeners everywhere are planting flowers just for the insects and adding bee hotels to their gardens.

What a wonderful shift! Job done. Book no longer needed. Well, not exactly. The choice of products to buy, organic or otherwise, to improve our gardens is huge and constantly growing. For every great new product there is more choice, not less. As gardeners we are bombarded with choices and products. We exist in a climate that is seeing a huge push on globalisation via social media. It is now easier than ever to purchase something from far away without a lot of effort or thought. If you have something to sell, you are told to think big and sell wide, even if your product is best suited locally. We are so brainwashed into thinking globally that thinking locally seems too small and insignificant, despite our awareness of environmental issues. We are now more environmentally conscious than ever and yet still so very far from combating climate change.

This book IS still needed. We may be gardening a lot more organically, but we are yet to really come to terms with gardening sustainably. Greater choice leads to increased consumerism. More watering of the lawn means more mowing, which in turn means more fertilising. We are still trapped in cycles of 'more'. A quick look at Amazon's top selling products in the lawn and garden section tells me this book is needed. Five of the top 10 were pesticides, for killing insects. Only one of these was organic. The top 10 also included two barbeque products, an inorganic fertiliser, a potting mix and a hose. Enough said, I have a book to write, and you have a book to read!

Let's start with a garden, or why we need more gardeners.

If you are reading this you are probably a gardener, or at least interested in gardening. Good on you, we need more people like you! Gardeners are pretty good people, sure, but that's not why we need more of you. Gardeners are more likely to be environmentally aware and make sustainable lifestyle choices. In the developed world we have extremely high resource-use lifestyles. We live in air-conditioned homes, drive air-conditioned cars to air-conditioned shops and air-conditioned offices. It is easy to see why reducing our energy use and thinking about living more sustainably might be difficult for us – our lifestyle has separated us from the natural environment. It is simply much harder to value something we have no connection to.

Gardeners are a little different. In most cases, our gardens are outdoors, in the dirt. They are living spaces – instead of being surrounded by manufactured objects, we are surrounded by living, growing, changing plants. We spend time outdoors, in the weather and noticing the weather, and noticing the living things around us. The simple act of caring for a plant is an act of environmental kindness.

In our cities and urban centres, population growth is putting pressure on available green space. We are losing green space to new developments at an alarming rate. We are also losing our gardens. Subdivisions into smaller lots with larger houses, and the increased use of hard surfaces in landscaping all contribute to less green space in our cities. In 2010, London lost an area of green space equivalent to the area of two Hyde Parks through the loss of private gardens – a significant loss not just of green space, but of gardeners too. Ten years later there is still a strong drive for development over green space. Meanwhile our New Zealand cousins in Dunedin contribute 50% of the city's urban green space through private gardens. That is a big contribution from a lot of people doing a little bit each. If you are in doubt about the value of your little patch of green, have a look at an aerial picture of your suburb online if you can, and then check back in a couple of years, or see if you can find an older aerial shot of the same view. Has the amount of green in the view increased or decreased?

These days urban planners are far more aware of the value of public green space and, although the exact value is not easily quantifiable, we are very aware that green spaces increase physical activity, help to create social cohesion and reduce stress and depression. Green spaces also reduce air temperatures, thereby offsetting the urban heat island, absorb pollutants from the air and mitigate storm water flows, which reduces the impact of flooding. They sequester carbon to some extent and have even been reported to reduce crime through creating healthier happier communities.

Phew! All that before we even talk about homes and food for wildlife, biodiversity or simple aesthetics. There are a multitude of studies to support all of these claims, but we need only look at the impact of a global pandemic forcing people into lockdown, to see how people craved the outdoors and in particular green spaces. I personally have never seen our local parks so crowded as they were during the 2020 COVID-19 pandemic lockdown, which is somewhat ironic. As the urban planners are increasingly valuing green space, it seems the average household is not. We certainly lead far busier lives than our grandparents did, and we no longer need to grow our own food, so the role of the garden has changed. Children seem to play outdoors less, and if they do it is likely to be at organised sports on public sports fields, rather than in the backyard. Recent studies have found that in Australia on average, prisoners spend more time outdoors per day than children do! I hope we have the sense to ensure that this changes.

How many of us remember our grandparents' garden? Back then they did not rely on books and television, nor did they have the same access to nurseries and specialists to help them know how to garden. Back then, part of leaving home to get married was taking a bunch of cuttings from their parents' garden to plant in their new bit of dirt and start their own garden. From there, they swapped cuttings with friends, family and neighbours. They knew what to do because their parents knew what to do and they had to help out, so picked it up as they grew up. Gardening used to be a basic life skill, much the way cooking and cleaning are. Sadly, much of that old knowledge has been lost and our children are no longer the children of gardeners. While you are reading this because you are a gardener, and your household all know the difference between a tomato and a pansy, I'll bet you know plenty of people who don't.

There are so many things we can do as gardeners to improve our green star rating, and most of the time we are very willing and even eager to do just that. I think this is because as gardeners we see the bees and butterflies disappearing, and we notice them come back when we stop spraying poisons. We are outdoors and notice the heat and lack of rain, then we see what a difference a shower of rain can make. We are outside as the sun is setting and notice the sunsets and the way the evening light highlights the colours in the flowers, and we never want to lose this.

Whenever I am worried about the state of our environment, I go and plant another plant. When someone admires my garden, I give them some seeds or cuttings. I hope to encourage and inspire them to also become gardeners. Wherever possible, I give plants as gifts, and take the time to help plant them as well to give them a better chance of success in the hands of non-gardeners. I hope that by encouraging more gardeners, I am helping to make their lives a little healthier and happier, but also that by becoming gardeners, they will become better environmental caretakers as well. All this is probably not surprising to you. But what is perhaps more surprising, is the amount of good you are doing for the planet simply by growing plants in your patch of soil, and how incredibly integral the soil is in this entire story. The first mission here then may seem simple but often is harder than it looks – create more gardeners. Our cities need their contribution to our green spaces, and our planet needs their awareness and appreciation, and it needs their soil repair efforts.

Sustainable Gardening

While creating a garden may be good for the earth, there are many gardening practices that also cause harm. The use of poisons is an obvious example. There is a widespread attitude that plants we don't know or want should be killed, which can be very damaging. Even organic gardening has a large focus on how to kill weeds and bugs. These organic methods are by far safer and have less collateral damage than are non-organic poisons, but it is still a focus on removing the unwanted instead of allowing the unwanted to fill a valid natural role.

Gardening practices can be filled with all sorts of other earth-damaging practices such as excess consumerism and packaging, depletion of natural resources, reliance on fuel driven tools and concrete. I am amazed at how much concrete is used in modern gardens. If we can take some of these damaging practices out of the equation, and still have an easy to look after, beautiful garden, we are winning, and the earth is winning too. While soil is the key to a successful garden and is also potentially the key to reversing climate change, learning to garden in ways which support healthy soil will be crucial. While I talk about many aspects of gardening in this book, the single biggest and most important aspect is the soil. The majority of sustainable gardening ideas all help contribute to good soil, so this is also a book about soil. But before I get completely lost in the story of soil, let's look further at sustainable gardening practices.

I would love people to make some small changes to the way they garden in order to be more sustainable. To that end I subscribe to the principle of good, better, and better yet. If you go to the extreme of absolute best practice - good for you, well done but you are probably a (wonderful!) freak. For the rest of us, there should be no pressure to enact every idea.

The information contained in this book allows you to make informed choices and is presented for the most part without judgement. Some choices may seem extreme to some but may suit you perfectly, and in other situations the reverse may be true. I don't want anyone to feel overwhelmed that there is too much to do to be good sustainable gardeners – there isn't, necessarily. Just the little steps that you find comfortable will contribute to an overall healthier, happier garden and planet – and gardener!

How sustainable is your garden?

How sustainable is your garden? Throw away the checklists and comparisons! They are broad generalisations and not a reflection of what true sustainability looks like in your own patch. Growing vegetables is not sustainable if you don't eat them, or if you live next door to an organic market garden, in which case you may be better off growing flowers to attract pollinators and getting your vegetables from your neighbours. Your sustainable garden is a personal journey and personal experience and should be a very rewarding one – for you!

Organic versus sustainable

Let me clarify the difference between 'organic' and 'sustainable'. An organic garden is one in which all of the products used in the garden were once living or are produced directly from once-living material. This relates largely to soil improvers (e.g., compost), fertilisers and pesticides. In an organic garden pyrethrum sprays are ok to use as they are based on a plant extract; neonicotinoids are not as they are chemically derived.

While organic gardening offers SERIOUSLY HUGE improvements over non-organic gardening in terms of healthy produce and a healthy environment, it is not inherently sustainable. Organic fertilisers are still manufactured products, wrapped in plastic packaging and transported (sometimes long) distances to where they will be used. When used excessively they can cause environmental damage the same way inorganic fertilisers do. So 'organic' is only part of the 'sustainable' story, albeit a very important one. Where those organic fertilisers are replaced with homemade compost and the manure of backyard chickens, that organic has now become so much more sustainable.

Sustainability is different for everyone

Consider which garden is more sustainable:

Garden A has a large percentage of native plants, includes herbs and some citrus trees, is neatly mulched with woodchip, and watered with reticulated rainwater.

Garden B has no native plants, an old and unproductive grape vine and mulberry tree, no mulch, and is watered only with town water.

Which garden is the more sustainable? If we were to use a simple checklist, we would easily rate **Garden A** as the more sustainable garden. It includes the must-haves for sustainable gardening – native plants, food plants, mulch and rainwater. If we dig a little deeper the picture changes dramatically. What about other issues like hard surfaces, use of chemicals, amount of new versus repurposed materials used, amount of petrol used in garden care equipment?

Garden A has a large concrete driveway, concrete and paved areas for entertaining, a pool, concrete fences and garden beds, small garden beds which are watered with rainwater on a complex automated system, mulched with artificially dyed wood chip mulch and the plants are highly treated with fertilisers and pesticides. There are no weeds as the regular gardener poisons these. The lawns are mown weekly, and a blower used to keep it all tidy.

Garden B is owned by a little old lady. She has two concrete tracks for the driveway with lawn and weeds growing between and around them. The garden is full of flowers; even the clover in the lawn is allowed to stay while it is in flower. It is crowded and the soil shaded. The flowers self-seed. The back garden has a concrete path and Hills Hoist but is otherwise somewhat overgrown. Weeds abound but no poisons, fertilisers or pesticides are used at all. The fence is an old chain link fence that is still perfectly functional after 50 years. Plastic cake trays from the supermarket are reused as saucers for pots.

These are real scenarios, based on actual clients of mine. I long ago walked away from **Garden A** as I found gardening in this way unsatisfying, but I will always help out the little old lady in **Garden B**. My own garden is an entirely different story again.

Sustainable gardening is far more complex and involves so many different choices than any checklist can really allow for. Often, we are told to use drought tolerant plants and recycled materials. That is fine if you live in an arid region (or are mid-drought) and have a recycling centre nearby, but for most of us, it is just not that simple. That oversimplified view can become very unsustainable when conditions change from drought to flood and the entire garden is waterlogged and needs replacing.

Nor is that all that sustainable gardening involves. There are many different ways in which we as gardeners can reduce the environmental footprint of our gardens. Reducing the need for supplementary watering and using recycled materials in landscaping are just two of many options available to us.

There are more ideas listed in the guide to follow. Not everything here will apply to every garden. Ultimately your garden should be a place of joy for you, not a burden, so while I hope you are able to incorporate some of these ideas to make your garden more planet friendly, **what the planet needs most, is that *you keep gardening*.**

Guide to sustainable gardening

- Match plants to the garden position that suits them best
- Encourage biodiversity of plants & animals
- Attract local wildlife & insects into the garden
- Reduce hard surfaces that cause rainwater to run off
- Don't let environmental weeds escape your patch
- Manage your lawn in a sustainable way
- Understand and care for your soils
- Use your garden to grow some food
- Use power tools less often
- Make your own compost
- Optimise plantings to reduce the need for electric cooling and heating in the home
- Choose garden features that run on renewables
- Consider sharing your pool with aquatic wildlife in a natural pool
- Adopt climate appropriate gardening, including climate appropriate plants
- Use recycled water or grey water as much as possible
- Use less pesticides & herbicides
- Use less fertilisers (or replace with organic ones)
- Source plants locally & choose some local natives
- Be a mindful consumer & garden ecologist

Consumerism

Consumerism is a huge part of what makes most of our lives unsustainable and gardening is no exception. We have all been to the large hardware chains and walked along seemingly endless aisles of gardening products. Many of us have stood there bewildered by choice and had no idea which product to buy.

There is no way our plants need all those products! Just as in every other industry, those products are sold by companies who are in business to make a profit. Some are more ethical and sustainable in their practices than others. We have the power as consumers to choose what we buy.

As consumers we are increasingly making conscientious choices about our everyday grocery items, but few of us have applied those same choices to our garden spending. Travel miles, excess packaging and provenance of the ingredients all apply to gardening products as well as to food. Conscious purchasing is part of the story. By simply buying less we are being more sustainable – less trips to the shops, less packaging, less waste, less energy spent on production etcetera, etcetera.

Go and look in your garden shed. How many products are sitting in there barely used? Or bought 'just in case' but never used? As a professional gardener I found the arsenal of garden products that was travelling from job to job with me was getting ridiculous. I had a polystyrene box in my car filled with the solutions to all sorts of garden problems. I had white oil, neem, weedkiller, pyrethrum, copper sulphate, fungicide, trace elements, iron chelates, lawn grub spray, nutgrass poison and an all-purpose fertiliser. My car became a mobile chemical hazard!

They all ended up in the bin and I took a fresh approach. These days the only product I purchase and use in both my own and clients' gardens is a very good quality rock mineral product which is high in silica and microbes and is locally manufactured. Without the chemicals, and with some simple adjustments to my gardening practices, all of the gardens I cared for became healthier places with happy thriving plants and no need for all those chemicals.

We will cover fertilisers, herbicides and pesticides in more detail in future chapters. These are all products that we tend to use in excess and which we actually need very little of. We will also look at power tools and plant purchasing in more detail in upcoming chapters.

If we think about the common sense of being a conscious consumer and apply that to gardening, we will immediately make an enormous difference to the overall sustainability of our gardening activities. These choices can be applied to everything we purchase, and the idea of sustainable choices when purchasing products will come up throughout this book. A few brief points to keep in mind:

Do I really need this product? Weed mat is a case in point. It reduces weed growth initially but over time makes no difference and causes depleted oxygen in the soil below it. We can achieve the same result with old newspapers or cardboard and save the planet from the additional plastic and production costs.

Is this product ethically sourced? (for example, recycled waste product versus intensely farmed or mined product).

Is there a homemade or organic alternative? There are very easy homemade organic options for nearly all pest problems, and even for dealing with most weeds.

Buy quality over quantity, in particular in relation to tools. Aim for a product that will last and does not need to be replaced often but also remember a high-quality expensive tool is no good if it doesn't suit your purpose, such as being too heavy for you to hold.

Avoid excess packaging. Many companies make a conscious effort to reduce or use recycled and recyclable packaging, but we can all do our bit as well by choosing thoughtfully.

Buy local to reduce transport miles. This usually means buying from a local company to reduce imports but could also be purchasing plants from local growers and suppliers. Congratulations, you are already thinking differently and making small but hugely significant changes to your gardening practice.

Some Gardening Basics

The following information is probably something experienced gardeners have a handle on already. As a beginner gardener you are probably busy learning all about plants. That is wonderful but before you try and create a garden you need to know this stuff.

There are three environmental factors that I would like to address before we look more deeply into sustainability. These factors are often put secondary in gardening guides but in fact, should be the very first thing any gardener tries to understand before getting too far along with a new garden – sustainable or otherwise.

These are:
- Understanding your climate
- Understanding your site
- Caring for your soil

It is important to give yourself the advantage of knowing your own gardening conditions and the factors that affect your block. For a garden to be successful it has to be appropriate to the climate it is in: you need to be aware of how the climate is specifically affecting the site it is located on (and therefore what the prevailing influences will be), and it has to be suitable for the soils that already exist on the site. By understanding these factors, not only will you end up with a more robust and successful garden, your garden will naturally be more sustainable as it is a better fit to its location.

Understanding your climate

How well do you understand your local climate? I am constantly surprised that people who have lived and gardened in one place for decades can so easily forget that it is very normal to have very little rain in Spring (for us here in the subtropics). If we know certain times of the year are dry, we can plan for that and water as needed, or choose plants that can cope with a drier climate. What climate zone do you live in? When is the wet time of year and when is it dry? Even within the broad climatic zones there are enormous differences due to proximity to the coast, or river or urban heat island, or elevation. If you live on a slope, is it north or south facing? Even your soil type will have an impact on what grows and that will impact your local microclimate.

Factors to consider are not just temperatures and rainfall. The following list of climatic factors are all important to consider when planning a garden with regard to what you can grow, what care it needs, and how prone your garden may be to drought or flooding. Surprisingly, very few gardening books cover this and yet it is so fundamental. I am constantly amazed at how often I am asked simple questions that relate to understanding the climate we are gardening in. This may be because I live and mostly work in a subtropical climate.

The majority of the world's gardening information comes from a temperate climate and needs to be adjusted to work here. Still, I have been met with questions such as "why do the ends of my frangipani branches keep rotting?" from gardeners in temperate Melbourne. The answer is because it is a tropical plant and the frosts in Melbourne kill the growing tips. It is being grown in the wrong climate. The world is getting smaller in terms of tourism, media reach and the ability to mail order plants from far away. Gardeners are plant lovers and we ALL hanker for something that won't grow where we live. We have all given something a go just in case. Rarely do these experiments work without a huge amount of care.

I love being a garden tourist, going places and seeing gardens full of amazing things I wish I could grow. I love being in faraway places and seeing plants that are super common and easy to grow at home being sold as rare. It is a great reminder to appreciate what I have at home. I do NOT love mollycoddling a plant and hoping for a result that is always disappointing. Over the years I have learnt that I am a far happier gardener when growing plants that grow well where I garden. The first step to being such a happy gardener is to understanding where you garden so that you can make better choices for your plants.

Climate zone

Broad climatic zone is a good starting point, but keep in mind that these are broad categories and don't always fit well. For example, technically the subtropics has hot humid summers and mild dry winters which do not drop below 6°C, and yet there are plenty of pockets within the subtropics which do not follow this general pattern, and get regular frosts when temperatures drop below 1°C. The US climate zones are often used in international garden literature. These, and the European and British gardening climate zones are all based on minimum temperatures. The US system has twelve zones. All of Australia fits into the four zones at the warmer end of this scale, and yet we know there are a lot more than four broad climate zones that impact on gardeners here. Broadly speaking at this stage, identify your location as cool temperate, temperate, warm temperate, Mediterranean, subtropical, tropical, wet/dry or monsoonal tropics or arid. You will notice from these descriptions that rainfall patterns play a major part in determining our climate zones.

Frosts

The minimum temperatures experienced will impact what you can grow. You should get to know if, and when you will get frosts and how severe they are. Frosts can be mild but persistent for many months, they may be occasional, or they can be deep frosts where the soil also freezes. Frost will govern your plant choices as different plants will tolerate different depths of frost, but it will also be an indicator of the chill factor in general which matters in order to get fruit or flowers for some plants.

Any of the stone fruit plus most berries, apples, pears, and so many other plants all require a certain minimum temperature in order to set flowers and fruit. Without this chill factor they may grow but not flower or fruit. Frosts can also be linked to humidity. Areas with high humidity are less likely to get frosts as the moisture in the air keeps the air warmer. Areas with extremely low humidity are also less prone to frosts as there is less available water to freeze, even if the temperature is cold enough.

The temperature at ground level can be a few degrees lower than the ambient air temperature, depending on what your surface is. This means you can easily get frosts even if your daily minimum temperature is 4°C.

Rainfall

This obviously impacts on drought/flood patterns. It also impacts on what you can grow based on when you get the rain. In the tropics and subtropics, the rainfall tends to be predominately November to April, making for hot, wet summers. In the temperate and Mediterranean zones, rainfall is usually higher in the winter and spring.

The time of year for the rainfall will help you know what sort of plants will prefer that pattern, and when you are likely to be most impacted by excessively dry or wet periods. Many tropical plants will enter a dormant phase in the subtropics or cooler zones. This is typically a dry time and therefore they prefer to be kept dry when they are dormant. If you live in an area with wet winters, you may be able to grow these plants, but they will need to be in pots that can be moved out of winter rains. Conversely grape vines are also dormant in winter, but they are a Mediterranean plant so are used to receiving water in winter. If you are growing them in a dry winter climate you will need to remember to water them even though they are dormant.

Humidity

Sadly, still largely left off the commercial nursery industry's radar when it comes to plant labels is humidity, which has a huge impact on what you can grow. Some plants will only grow in high humidity and others will only grow in low humidity, so it certainly matters to plants! Humidity impacts on transpiration and evaporation rates and will impact on how plants cope with dry or wet conditions.

Areas that typically receive their highest rainfall in summer are likely to have high humidity. Areas that receive their main rainfall in winter are likely to have low humidity. Plants that are adapted to low humidity conditions tend to have small leaves that are hard, furry, or grey, or a combination of these. It is an adaptation to prevent excess water loss through transpiration. In highly humid conditions these plants tend to suffer fungal diseases and can die quite suddenly, which is what lavender tends to do when grown in the subtropics.

Plants adapted to humid conditions tend to have large leaves with a large surface area to facilitate transpiration of excess water. The leaves may also have a drip tip; a pointed tip which droops slightly, encouraging excess water to run off the leaf. In low humidity these plants wilt excessively and keeping water up to them is an endless task. Many plants fall somewhere in the middle of these extremes and will have some tolerance either way.

Wind

If your garden is exposed to high winds it will tend to be a lot drier. If those winds are coastal, they may also be salt laden. The direction of strong winds and storms will be important in determining what to plant where in your garden to minimise wind and storm damage. Wind will mean lower humidity (perfect for growing lavender in humid climates), but constant strong winds will mean lots of drying out and will also put pressure on the plants to anchor themselves securely to resist blowing over.

Large leaves tend to get shredded by wind, so a windy garden will be one for small leaves or those with an open structure such as palms. Even if you do not get wind as a regular factor, you may get regular storms, which usually bring strong winds with them. Knowing the direction of prevailing winds will allow you to find the more exposed or sheltered parts of your garden and to choose appropriate plants for those areas.

I had to give up on growing bananas on my southern garden boundary. It was a sunny spot with plenty of water but when the summer storms rolled in from that direction, they constantly shredded the leaves and blew over the fruiting trees before the fruit was ready.

Elevation

Often you will have cooler, less humid conditions at higher altitudes, with higher rainfall. The southern side of a mountain will generally be cooler and shadier so not dry out as quickly as a north facing slope (in the southern hemisphere that is!), which can impact not just on drought and flood issues in the garden, but also on the predicted path of bushfires. The coastal side of a mountain range will usually receive higher rainfall than the inland side, and the position of hills and mountains can create local rain shadows. Elevation could also mean exposure to winds.

Slope

A steep slope may have trouble retaining moisture, making it more susceptible to drought. It may also be highly susceptible to erosion during high rainfall events. At the opposite extreme, completely flat ground will drain more slowly and could be more prone to waterlogging.

It can be very difficult to get water to penetrate into the soil on a slope, instead of simply running off the surface. While a densely planted slope can help slow water flow, certain plants are more useful than others. Dense strappy plants such as lomandras are fantastic for creating a physical barrier which slows water flow. The dense fibrous roots are not easily eroded and help to stabilise the slope. If possible, dig some trenches across the slope just above the position of new plantings. The trenches don't need to be deep, 20cm will work if you can dig that deep. Fill the trench with organic matter of any description, then put the soil back on top. Water will drop into the trench as it runs down the slope, and will soak into the soil from within the trench. The organic matter will compost in the trench, helping with soil improvement. As the organic matter composts, it will drop, leaving a dip or swale in the soil which will continue to slow the flow of water and ensure water is soaking into the soil.

Land Use

Local land use will also have an impact on your climate. An area with significant forest cover will tend to have a milder climate than one with open fields. If you live in an urban environment, your local climate will tend to be hotter and drier (known as the urban heat island) due to the abundance of heat absorbing hard surfaces. The urban heat island effect can make a very significant difference to the climate in your garden and should not be underestimated.

My own garden is in a subtropical climate (Brisbane), in an inner-city suburb and in a natural dip in the land. I experience very little wind at ground level and so have quite intense humidity as it can sit in the low land. Winds can be turbulent during storms as there is a lot of infrastructure around for wind to bounce off in terms of nearby developments. The area is highly developed which caused temperatures to increase, being particularly noticeable in winter as winter lows are not so low as they once were. The low ground has restricted drainage and tends to water logging during wet weather.

All of this has very much affected how I garden and what I can grow. I have had to install drainage trenches and pits under my garden to prevent flooding. I have also done huge soil work to increase water penetration of my heavy soils. My garden is on the edge of a flood zone, so managing water is critical here. Many of my gardening colleagues live in outer suburbs with very different local climate influences, so while we are all subtropical gardeners, and even all Brisbane gardeners, we do garden under quite different conditions.

Understanding your local climate will give you a lot of information, not just in terms of managing water in your garden, but also what sort of plants to grow, how to structure your garden, what time of year various garden tasks are best done and what natural risks your garden is likely to be exposed to. It will be your guide to how you approach setting up and caring for gardens, what inputs may be needed to begin with and on an ongoing basis, and this in turn creates an opportunity to make informed choices which will not only improve the overall success of your garden but also the sustainability of your garden.

Understanding your site

Once you have a good broad understanding of your climate zone, you will need to apply this to your site and look more closely at the microclimates that exist for you. The more you understand this, the better the plant and landscaping choices you make will be. Factors affecting your microclimates will include:

Aspect

Locate the compass points so that you know the trajectory of the sun across your garden throughout the day. Ideally a year's worth of data would be great as the sun is lower in the sky, the shadows deeper and even in different places in winter as compared to summer. Realistically, very few of us are going to wait a year before starting a garden, so you will need to keep on watching and learning as you garden. In a southern hemisphere garden, a south facing garden will be mostly shady, especially in winter, and a north facing garden will be sunny all of the year. This is reversed in the northern hemisphere.

Sunlight

Do you have trees, buildings, or fences blocking vital sunlight or creating welcome shade? Can these be manipulated for greater garden success, or do you need to understand how best to garden within the constraints of these conditions?

Exposure to indirect sunlight and heat

Light from reflective or light-coloured surfaces can be very beneficial in dark or shady areas, but hard surfaces can also store and reflect heat that can make areas of your garden exceptionally hot and harsh. Be aware of the impact that brick or concrete walls, foot paths, paving and even gravel surfaces can have on heating areas of the garden. Can you soften the impact with plants, or use this extra heat to your advantage? Large open areas of lawn can also become very hot areas, which may benefit from a little shelter from a well-positioned tree, although they will be cooler than the same area in any non-living material such as fake grass or concrete.

Slope

Look at where rainfall runs away and where it pools. It is easy to initially think you have a flat garden but looking closer will reveal the local contours. Do you need to redirect rainwater away from the house? Do you have a low corner where puddles form? These small natural contours will contribute to differences in soil moisture. By noticing them it will help you identify the wetter and dryer parts of your garden. This information can help you make important adjustments to drainage, or choose the best planting spots for fussy plants.

Airflow

Your garden may not be in a windy location but there will be places that get more free airflow than others. Airflow will relate directly to humidity so when you understand this it can be used to your advantage. Airflow will be greater in the more open parts of your garden, but you may also find good airflow along the front fence line where breezes follow the street. I have clients who grow amazing lavender in Brisbane, a classic climatic mismatch. One has the lavender growing on a west facing rock retaining wall above a wide street. Another has it growing on a rock retaining wall facing the river. In this case the constant slight breeze above the river keeps the air moving enough to reduce humidity. Neither of these people can grow lavender in other parts of their garden.

Such a detailed understanding of your site may not be evident at first but become more apparent over time as you watch the garden grow and change. Trees will grow and provide new shade. Those trees may also create dry zones by sucking all available moisture from the soil. Neighbours might cut down trees that were overshadowing your garden. The better you understand your site, the more you will grow with your garden. After all, gardens are not an end point but a constant journey.

Get to know your soil

Your soil is your greatest asset, so know it, nurture it and protect it. Strong plants start with healthy soil, but before you can make your soil healthy you need to know what you are starting with. A good understanding of your soil will help you know how to manage it, how to improve it and what is likely to grow well in it.

There are soil maps on the internet, but they tend to be rather technical. To begin with dig a hole and see what you find – there is a lot to be learned by digging a hole and having a look inside. This is not a book about soil types, as this is an extensive subject in itself, but if you don't know what you are looking at the following information will help.

The two broadest distinctions are clay-based soils and sandy soils. The most desirable soil lies somewhere in the middle of these two extremes and is called loam. To find out what you have, take a handful of soil and wet it until it becomes moist but not sodden. Squeeze it together to form a sausage shape. A clay soil will hold its shape easily and will have a tendency to get sticky.

A sandy soil will fall apart and not hold any shape (you may also be able to see grains of sand in it). A loam soil will hold its shape to some extent but will remain crumbly.

You may also have a silty soil, characterised by fine particles which feel chalky to touch. This sort of soil can form a hard crust when it dries out. For a gardener, soil can vary enormously even over a small geographical area. Here are the basic questions you should be investigating before you get too far with gardening:

Soil type – is it sandy, clay-based, or a nice friable loam?

Topsoil – how deep is it? Good topsoil should be friable (easy to dig) and have plenty of organic matter and soil life (worms and other microbes). Rich deep topsoil is a gardener's dream. This is the layer where all the action is, where the worms and microbes are busiest, and where the plant roots are. Shallow topsoil will result in shallow rooted plants, which can be very problematic.

What is under the topsoil? – how far do you need to dig to hit compacted soil below, solid clay, rock, or even the water table? Trees of course need to be able to get their roots deep into the substrata below the topsoil in order to get sufficient anchorage. If your substrata is rock, or solid clay, this can pose problems which will need to be addressed if you wish to grow trees.

Understanding what lies below will help you to know what to expect. These days what lies below is often on top. In building new houses, earthworks are usually careless in protecting the soil. The clay that is dug out for the foundations gets dumped on the surrounding surface and the new owner is faced with creating a garden in ground which may be full of builder's rubble and heavy clay.

Regardless of what your soil is, or how deep it is, you will be far more successful in creating a garden if you can identify what type of soil it is. Understanding the soil is not just about which plants will grow there or what soil preparation may be necessary, it is also about understanding the impact of water in your garden. Heavy clay soils or soils with a shallow clay substratum will easily waterlog. Sandy soils and soils with rock below will be very free draining and may have trouble holding moisture.

The Kerkin Garden - A tale of many microclimates

John and Carol have been gardening their large semi-rural garden for 40 years now. They have the advantage of maturity of design and of the many trees they have planted, but for these avid plant lovers, there will always be something new to add, to create or to grow.

These two really are "shakin' what ya mama gave ya". In this case of course, mama is Mother Nature. The garden is set on an eight-acre property which was part of John's family dairy farm, and the place he grew up. When John brought his new bride here to live 40 years ago, it was at the end of a dirt track, and there was no electricity. Luckily Carol was not easily put off and together the couple have created a truly amazing garden in a beautiful rural setting.

The property is on a fairly steep northeast facing hill, with a creek at the bottom of the garden. The hill continues to rise behind the property and another hill rises in front of the property.

While these hills provide a magical borrowed landscape, they also contribute significantly to the climate impacts on the garden. In winter the sun doesn't rise over the first hill until approximately 8am, and will set behind the garden by 5pm, making winter days shorter and cool periods longer than the local average.

The garden is located in the Gold Coast Hinterland, and while this is officially a subtropical climate, the garden is cool enough to grow many plants that wouldn't normally cope in this climate. Things like magnolias, which are thriving here. A magnolia hedge (Michelia 'Bubbles') creates a wind break and privacy screen for the rose garden. A large *Michelia doltsopa* 'Silver Cloud' creates a heavenly scented overhead display and shelters the fairy garden. A *Magnolia soulangeana* creates a stunning winter focal point for the back dining area. But their favourite is the yellow magnolia 'Elizabeth', a birthday present from Carol to John.

The garden occupies approximately three quarters of the eight-acre property, so there is a lot of space to fit in favourite plants, but there is also a lot to care for. John points out where he has added some more Louisiana irises in the rose garden. There is a natural spring just above this area, so he is making the most of the extra water available in that spot by growing water loving plants there. Just behind the rose garden is a north facing rocky outcrop. It is hot and dry, the perfect place to grow aloes and succulents. A huge 'Eric the Red' aloe dominates the garden and puts on a dense spectacular flowering which invites you out of the rose garden to explore what is beyond.

Throughout the garden plants are located in spots which best suit their needs. Finding the many microclimates and matching the plants to these spots in the garden has been key to the success of the garden. For John, working with the natural microclimates in the garden is a no-brainer. As he says, "if you don't, you are working against yourself".

John and Carol do all the work in this huge garden themselves, plus John works full time, so there is little opportunity to fuss over plants that are struggling. Natural advantages need to be exploited wherever possible. Sometimes plants land in the perfect place by accident. The cymbidium orchids didn't flower in the garden by the house, and there was no way Carol was going to fuss over them by giving them ice water regularly, so they got planted onto a rocky patch by the creek where nothing much was happening. Here the foliage could look great, and it didn't matter if they flowered or not. Turns out, it is the perfect spot, they flower well down there. Being on a hill, cold air sinks and the plants needing cold spells do better at the bottom of the hill.

They have bulbs, perennials and annuals amongst the roses, many of which don't do so well higher up in the garden, or elsewhere in the subtropics. Care however has to be taken with pruning the salvias along the front fence line. If they are going to get frosts, this is where it will be, so pruning here has to wait until the risk of frost has pasted for the year. Elsewhere in the garden, the same salvias were pruned weeks earlier.

The garden is neat and tidy, which is a bit surprising for such a large garden. Lawns are mown weekly, edges and hedges are trimmed often. This is not the huge job that it sounds, as by doing it often only a small amount needs trimming, and it is a relatively quick easy job. It is a job John enjoys as he finds the outcome so satisfying. Certainly, there are no bare patches in his lush lawns and hedges. The 'little bit often' approach to both results in thick strong plants that always look great.

Pruning can be a big job, as this garden is overflowing with fabulous perennial plants. Carol always has a huge vase of flowers from the garden inside the house. "Chop and drop" is Carol's mantra. Even sticks are incorporated into the chop and drop as they help to create air pockets in the pile of green matter which enhances the composting process.

This process of chop and drop is their main form of soil improvement (together with rock minerals) and the annual prune then becomes an annual dose of compost. Even palm fronds are cut up and laid down as mulch in the rainforest garden, where they break down reasonably quickly in the damp environment.

John used to work in an engineering workshop, which he put to great use making amazing metals sculptures for the garden. These days he works full time as a gardener in other gardens, which is also a huge advantage for this garden. He brings home green waste from other gardens (so long as it does not contain cat's claw creeper or madeira vine) to add to their chop and drop composting, and he brings home cuttings, unwanted plants and even unwanted statues and feature pots that clients are getting rid of. They all find a new life in this garden.

The majority of this garden relies only on rainfall and gets no supplementary watering. Even the rose garden is only watered in times of extreme dry and when they are having an open garden. This is possible because they have matched plants to the right spot in the garden and put in extra soil care.

The garden nestles beautifully into the slope it is on. There are no large terraces and retaining walls. Instead, the garden works with the landscape. Gravel paths meander through the garden and gently up the slope. There are steps in the steeper parts. Most of the gardens are edged with rocks, all found on the property (mostly while they are digging a hole to plant something!). In some places the rocks have been built up into a small retaining wall, which looks fantastic and natural. This allows for small flat areas to be created into garden rooms.

You can move through this garden from room to room, with incredible vistas and views at every turn. John has been so creative in using old gates, arches and shrubs to frame views and to draw you from one area of the garden to another. In the various rooms are different seats and table settings inviting you to linger, or a solar fountain in a large water pot, or a feature urn.

In the fairy garden is an old swing set that was Carol's as a child, made by her grandfather. It is now used by their grandchildren. The pathway here has mosaiced stepping stones that their daughter made when she was a child. This garden has layers of family memories and stories.

There are old cream urns in the garden which still bear the family name 'Kerkin' from when this was a family dairy property. The old banana packing shed, from the days when it was a small banana plantation, is now the garden shed. It is full of rustic charm, not by design, but by its nature and the amount of time it has been there.

There are even stories to go with the huge ancient jacarandas at the front of the garden. Whilst working in the front of the garden one day, John noticed an elderly lady standing outside. He asked "Are you admiring my jacarandas?" to which the lady replied, "No, I'm admiring MY jacarandas". She recalled bringing home the seedlings in tin cans and planting them there when she was a child. Those jacarandas are now huge, and much loved.

Many years ago John and Carol attached stag horns to the trunks, thinking they would look good as they turned into the driveway. The staghorns all slowly died, but nature had other plans. These days the opposite side of the tree, which you see well as you drive out of the garden, is covered with huge staghorns, all planted by nature. This is the side of the tree where the prevailing winds were able to blow in spores, and as the rain also blows in on this side, they are watered sufficiently by nature to thrive.

The wildness of nature here is softened by the lawns, clipped hedges and neat gardens. Other parts of the garden are more wild. Everywhere there is the constant sound of a multitude of different birds and the buzzing of bees. The garden is a healthy ecosystem where no pest control is needed, apart from an occasional dusting of derris dust on crinum grubs. There must be holes in the leaves but there is so much else to see that I confess, I didn't notice any.

For both John and Carol, this garden is not work, it is absolute pleasure. They will be out in it till dark most days, even after a full day of work, which these days for John, is gardening. Their love of gardening has become their profession, and they couldn't be happier.

Carol propagates madly, providing additional plants for the garden and also for her small nursery business. So much of what is in the garden is the result of years of work and propagating. Even many of the roses were grown from cuttings.

But for all that, John loves nothing more than being down on his hands and knees weeding or pruning quietly and listening to the buzz of a bee nearby. The world really can disappear when you are a gardener!

Sustainable Soil Care

Good gardening starts with the soil, so too good sustainable gardening needs to start at soil level. I grew up gardening in deep sandy soils. These were hungry soils that seemed to have an insatiable appetite and thirst. During drought they were challenging to keep moisture in, even with decades of organic matter being added. During wet years they became waterlogged and boggy because it turned out the water table was not very deep there. Before clearing for development this land was marginal swampland characterised by lots of paperbarks.

Now I garden on a small suburban block that has in the past been quarried, then filled, then capped with clay. Once again, I am dealing with waterlogging but at least now it holds water well in times of drought. Many years of soil amelioration has made enormous improvements to both gardens. Whatever we may have, soil improvement is probably a good idea. I rarely come across gardens that naturally have rich deep loamy soils.

The addition of organic matter is the primary means of improving soil, whether that soil is sandy, rocky, silty, clay or any mix of these. By increasing organic matter in soil, we increase its ability to both hold moisture and to drain more freely, making the soil less susceptible to both drought and waterlogging. Organic matter also serves to release nutrients into the soil. It does this through not only releasing the nutrients contained in the organic matter itself, but also in absorbing and then slowly releasing any fertilisers that have been applied to the soil.

The importance of organic matter in healthy soils cannot be overstated if we wish to be successful and eco-conscious gardeners in the face of unpredictable climates. You can almost think of organic matter as a super sponge in the soil. Organic matter holds moisture and nutrients in the soil, therefore soils with good amounts of organic matter tend to be more drought resistant, more fertile, and grow stronger plants. Regardless of the problems you face in your soil, the answer most likely will be to add organic matter.

Organic matter in soil

Australian soils tend to be low in organic matter – often in the order of 1-2%. Most soils around the world have around 3-5% organic matter. In a garden situation we have high expectations of our plants in terms of faster and denser growth than would occur naturally. To sustain this faster growth, we usually turn to fertilisers – highly manufactured products. The use of fertilisers can be greatly reduced and even eliminated completely by ensuring soils have sufficient organic matter.

Interestingly, the huge forests of Fraser Island off the Queensland coast are growing in pure sand. Fraser Island is the world's largest sand island. The water table is not deep as trapped layers of leaf litter form barriers within the sand, trapping water and nutrients and leading to the formation of many beautiful lakes in the almost pure sand. This results in the perfect balance of available moisture and free drainage to allow these spectacular forests to thrive. Nutrient availability is low in sandy soils, but the constant leaf drop creates a moist humus layer, which is constantly breaking down and feeding plant growth. Many of the world's rainforests occur on naturally poor soils. The constant moisture and warm temperatures allow fast nutrient recycling in the humus layer, which is sufficient to sustain rapid growth. This has many implications for us as gardeners in the way in which we manage soil.

In most natural ecosystems input and output are in balance. Leaf production draws nutrients and organic matter out of the soil. Leaf drop creates a humus layer, which breaks down to return nutrients and organic matter to the soil where it is available to be taken up by plants to feed further growth. In many of Australia's dry eucalypt forests this is an inefficient process and dry plant material builds up, becoming fuel for forest fires. The fires, if they are not too hot, act to break down the fuel and return the nutrients to the soil in the form of ash. As we have recently seen, in times of drought in combination with climate change, these fuel loads can build up to such a point that fires are catastrophic. At this level fires are not recycling nutrients at all and are inflicting nothing but destruction. In the humid subtropics and tropics, the increased moisture and temperature greatly accelerates the natural breakdown process, making it a very efficient way of recycling plant material to build soil and feed plants. In cold climates the annual snowfall traps fallen leaves under moisture helping them to break down. Frosts also play a role in breaking down leaf matter and returning it to the soil. The act of freezing breaks down cell structure, as anyone will know if you have taken vegetables out of the freezer and let them thaw before using them.

In gardens we tend to constantly remove plant material, creating an imbalance. We address this imbalance by adding fertilisers and mulches. In the average garden it is not possible to grow a rainforest with an inbuilt recycling system capable of handling massive amounts of green waste. We can however operate a similar system on a much smaller scale by simply viewing our own green waste as our primary resource for soil improvement. In short, we need to emulate nature by balancing inputs and outputs of organic matter. Whatever you prune from your garden is organic matter drawn up from your soil. The more you remove this green waste from the garden, the more depleted your soil is becoming, and the more work you will need to do in order to top up this organic matter and increase your soil fertility. If you don't take it away, you won't have to replace it.

In its simplest (and laziest) terms we can practice chop and drop gardening. This option is my favourite as I am time poor in all the gardens I manage. Simply let your prunings fall into the garden bed and break down right there on the spot. Weeds also fit the bill here, even if they are seeding. Once the plant has set seed, there will be a seed load in the soil, which you will have to come back to regardless of whether there are five or 500 seeds. A few more will make no difference and will reduce the risk of spreading the seeds elsewhere. Chop and drop most closely replicates a natural system. Most soft prunings will break down fairly quickly in situ and act as mulch on the garden as they do. The downside is that it can be unsightly.

One rather large garden I worked in had shrubbery which hid most of the ground. Here the client followed my advice and used all prunings and weeds as mulch. After doing this consistently for two years the condition of the soil improved enormously, from rather light grey to a rich dark brown that is full of worms. We also added some horse manure and pelletised poultry manure but as access to this garden was difficult, removing green waste and bringing compost and fertiliser in was a painful task. Recycling green waste in situ was a great time and energy saver, which had the added benefit of saving money and resulted in massive soil improvement.

A shredder can be used to put larger prunings through before dropping them on the garden. It turns your prunings into homemade mulch. This is also a good idea for tough leaves such as eucalypt leaves, which don't break down easily, or if you have a fairly dry climate that slows the breakdown process. It can look a lot neater to chop your prunings up before dropping them on the garden. Having said that, it is amazing how much you can hide under shrubs! One of my workshop participants some years ago was a commercial gardener. I was thrilled to hear that after attending a workshop and learning about chop and drop gardening, she now practices it in the gardens she cares for. Not only has she saved herself a lot of time and cost in disposing of green waste, she feels great about the fact that she is feeding the plants in the process. As she hides the prunings under shrubs, no one even knows that she has dramatically altered her gardening practices.

There is a long-standing school of thought that by doing composting in situ you are encouraging disease in the garden. This is rare but can happen. As we will discuss later in the book, good soil care and sustainable gardening practices can massively reduce disease risks in the garden, however if you are concerned that you may have diseased plant material, it is best disposed of carefully. If in doubt, call in a garden consultant for advice.

Mulch or Humus?

At this point let me explain the difference between mulch and humus. Mulch is a layer on top of the soil that acts to protect the soil – from water loss, temperature extremes and erosion. Mulch can be organic or inorganic.

Humus is the thin layer of organic matter that occurs at the surface of the soil. It can be the layer in which organic mulch is in direct contact with the soil, or it can be the layer of compost or prunings below the mulch. Humus is always organic and it is where the greatest amount of activity is happening in the soil. The humus layer will include fungi, bacteria and other microbes in addition to many tiny invertebrates that are all working to break down the organic matter and incorporate it into the soil. The action happens at the surface of the soil, so layers of organic matter can act as both humus and mulch.

Organic matter is important in soil for many reasons. Organic matter:

- helps break up clay soils, increasing the porosity (the open spaces in the soil) of these soils so that water can more easily penetrate rather than sitting on the surface.
- binds sandy soils, increasing the ability of the soil to hold moisture and nutrients.
- increases the water holding capacity of soils, which is critically important in preparing for drought or extended dry conditions.
- adds nutrients to the soil.
- absorbs fertilisers and re-releases them slowly as plants need them.
- feeds microbes and other soil invertebrates which in turn help keep the soil aerated and convert nutrients into plant available forms.
- buffers pH extremes.
- helps the soil to be more resilient to pollutants.
- stores carbon in the soil.
- reduces the chances of the soil becoming hydrophobic (when water runs off the surface leaving the soil completely dry below).

There are many ways of adding organic matter to the soil, from leaving prunings to rot in place, natural leaf drop, animal manures, organic mulches, mushroom compost, organic based pet litter products, compost, seaweed, grass clippings and even ash. The most sustainable source will always be the most local and nothing is more local than homemade compost made using your own green waste. Commercial composts and manures involve transport miles and packaging, but as they are turning a waste product into a valuable resource, they are a good option if you have no other source of organic matter.

There is a school of thought that tells you to dig compost or organic matter in to the top 30cm of the soil. I am not from this school and prefer not to till the soil at all if possible. That is not just out of laziness, although why do more back breaking work than you really need to? By breaking up the soil you damage its natural structure, which can be more delicate and complex than we fully appreciate. It will also make the soil more susceptible to erosion and kills worms. I find there is rarely any need to turn soil (except when planting). I prefer to add organic matter to a decent thickness – at least 5cm, more if the soil is bad and the garden bed is fallow, and let the worms do the rest. I have added aged stable manure in a layer 5cm thick over an entire garden, with a light sprinkle of woodchip mulch over the top and a handful of chook poo pellets underneath, and found that there was no sign of it one month later when I went to put more plants in. The soil however was darker and moister, and had a lot more worms in it. I do this in many of the gardens I care for in the early months of summer to help prepare the soil for what may be a hot dry summer. The difference is very apparent; however, this amount of organic matter will need to be reapplied at least twice a year for a few years to get long term and deep improvement of the soil.

In general, most organic matter can be added to the soil by simply layering it on top of the soil and letting it become humus. Nature will do the work for you. Mulching over the top of the humus organic layer will help to keep it moist and therefore break down faster.

If you have a large area of fallow soil (soil without plants in that you hope to plant in coming seasons), you can add the organic matter very thickly to get a bigger initial input of organic matter into the soil before starting your garden. You can even make this area a very large compost heap – throw your kitchen scraps, cover with leaves, manures or grass clippings and once it has rotted down and is ready for planting, you will have greatly improved soil.

When I initially moved to my current home, we found the soil was worse than bad. It was very heavy clay and the lightest shower of rain simply lay on the surface as puddles. As there was no garden here at all, I had a clean slate. Although I was desperate to start gardening, without attention to the soil, it would never work. I needed to do a huge soil improvement job as quickly as possible. I organised to have grass clippings delivered for free from the local mowing contractor. I had huge steaming mounds of grass clippings all over the place, all at least a metre deep. The heat killed off any seed so weeds were not a problem. Our local mowing man was also a fairly conscientious worker who kept the lawns he cared for regularly fertilised. This meant I often got clippings which included the first flush of growth after fertilising, so were extra high in nitrogen (another good reason to keep your own green waste). Actually, he was always a little bit gleeful when he dropped me clippings from a recently fertilised lawn, as he, a retired telco worker turned mowing man, could see that the constant removal of grass clippings and addition of fertilisers was a wasteful one-way street. It is thanks to this initial input into the soil that I now have decent topsoil. If I had my time again, I would have been more patient with starting the garden and added much more organic matter early. Now that my garden is full of plants it is much harder to add organic matter. When a planting space appears, I add aged horse manure or compost from the chook pen. Otherwise, I practice chop and drop and as many prunings as possible are tucked in under the bushes to replenish the soil from there. In areas that are largely self-sustaining and need very little pruning, their own natural leaf drop is the main form of organic matter this garden receives. Be aware that kitchen scraps, fresh manures, and fresh grass clippings should all be composted before adding to the garden. They will compost in situ but can hurt the plants in the process so either pre-compost before adding to the garden, or only use them on fallow beds.

Once you have added all this organic matter, it is time for a cuppa. Be patient and let the worms do their job. Soil improvement is not an instant process. If you are adding good quality compost that is fully composted, or even aged stable manures, it is mild enough that it will not hurt plants and can be planted straight into. Other organic matters such as aged manures, seaweed, mushroom compost, organic pet litter (remove the solids first, a bit of cat poo won't hurt the garden but it does contain pathogens which can be hazardous to human health, so best disposed of in the bin or the toilet) or leaf litter is fine to scatter around an established garden bed regularly, even a few centimetres thick, and it will give ongoing continuous improvement to the soil. Organic matter does get used up in the soil, so replenishment over time will be necessary.

Organic matter that is high in salts or nitrogen should be used sparingly, or in combination with other organic materials. Raw animal manures that are high in nitrogen such as poultry manures and even cow manure can burn plants. Manures such as horse, sheep, goat, rabbit and kangaroo or even alpaca tend to be fairly safe and mild as they are largely just digested grass. Even still, fresh can burn in addition to being smelly. The smellier it is, the more it needs to be composted first. Your nose is a good guide as to what is safe to use in the garden – the more pungent it is, the more it should be composted first.

A note about stable manures: there is much talk about the chemicals given to horses, especially racehorses, and how these impact on the safety of the manure. If you are trying to grow organic vegetables, you will need to make sure you only source organic composts, and this does mean any manure-based products need to come from chemical free sources. This can be difficult to do as we generally don't know what the farming or animal husbandry practices might be unless they are specifically labelled as organic. The worming medication given to animals can have an impact on the manure and many gardeners will not use horse manure particularly as they are worried it will kill worms. The reality is that worming medicines break down quickly. If the horse manure is aged, it will be not only safe for worms, they will love it.

One other reason to always used aged manures is seed load. Horses do not digest their food particularly well and a lot of seed comes through safe and well. When the manure is aged, much of the short-lived seed has passed its expiry date. The longer-lived seed can still be there and ready to sprout. This is actually what I love about horse manure. They have pretty good taste in greens, and so very often the weeds that do grow are ones I want. They are good edible greens such as mallows, fat hen, chickweed, nettles and more. These weeds are great food for us, but also make great compost, so if I can't persuade you to enjoy them in the kitchen, harvest them before they reseed and add them to the compost. A layer of mulch over the top of the horse manure will not only prevent any seed from growing, it will speed up the composting process.

There is a way that you can use some of these products 'in-the-raw' so to speak in the garden. Bury them! Dig a hole or a trench and bury your uncomposted organic matter, backfill and plant on top – into the soil on top that is, not into the compost heap below ground. I really like using this method when gardening on a slope, as it is very effective at sucking the water in and preventing water running off the surface of the slope. If you don't mind digging, burying organic matter is a very effective way of getting it to compost in situ, break down quickly and attract worms to your garden beds. All manner of once living materials can be buried for soil improvement. I did learn a lesson about the depth of the hole the hard way. I decided a fish carcass would make great compost (which they do!), but alas I did not bury it deep enough. Our dog was straight over the fence, dug it up and rolled in it. That hole was not nearly deep enough! As a rule of thumb, aim to bury your organic waste at least 30cm deep, especially if you wish to have soil space to plant directly above it. There is another advantage for burying organic matter deeper in the soil. The deeper it is (within reason of course, too deep and it is outside the active zone and nothing will happen), the deeper you are creating an alive and fertile zone in your soil, thus encouraging your plants to send their roots deeper into the soil in search of the good stuff.

Mushroom compost can have a very variable pH, although it is often highly alkaline and can be high in salt, so it should always be used in conjunction with a good organic mulch, or added to the compost heap first, and never relied on as the sole source of organic matter in a garden. At the opposite extreme, coffee grounds used too heavily can add acidity to your soil. Diversity of organic matter is always a good idea.

Seaweed does not need to be washed first as it does not have a huge amount of salt on it, and it is unlikely that you will have enough of it to cause a salt build up in your garden. Seawater is not just salty, it is rich in minerals, so the residue on the seaweed can also be rich in minerals. Be aware that many councils have limitations on how much seaweed you are allowed to harvest under normal circumstances. Every so often weather conditions cause huge loads of seaweed to wash up on the shore, becoming a nuisance and a hazard. This is the perfect time to collect it.

Salt problems are more likely to come from using overly rich composts in large amounts, or excess fertilising. In areas with high rainfall salt and nutrient build up is less likely to be an issue as the high rainfall will leach these out of the soil. In fact, in the wet tropics the opposite is a problem. By the end of the wet season, soils can be very depleted as the rain has washed so much goodness out of it. Salts and nitrogen are both highly soluble, so if they do build up in your soil, they can be flushed out with water. Being soluble, this water is now laden with those salts and nutrients so you need to be mindful of where it ends up.

Compost that is too rich is one that is very high in manures or nitrogen based raw materials. These raw materials need to be balanced with some 'brown matter'. In compost terms, brown matter could be dry leaves, shredded paper, woodchip, wood shavings or straw – materials with a high carbon content. Getting the balance right in a home compost is easy. If it is too wet and smelly, add some dry carbon materials. If your compost is too dry, add some wet matter which is higher in nitrogen such as kitchen scraps, lawn clippings, manures or even wee on it yourself if your compost is not too public. There seems to be a huge and complex science being applied to compost. I have even been approached by a company wanting me to endorse a smart phone app to monitor home compost for temperature, moisture and carbon to nitrogen ratios. This is all fantastic if you are making enormous piles on a commercial scale, but for the home gardener this is taking it to extremes. While there are some aspects of the science of composting that do help, like sitting your heap or bin in a warm place, and aeration to speed up the process, the basics of balancing wet and dry ingredients is the fundamental of compost making. This does not need to be measured as simple observation will tell you what is needed.

The role of soil care in earth repair

All that soil work is critically important to us as gardeners, but how does this save the planet? Well hold on to your hats, this is where it gets really exciting.

As it turns out, soil with higher amounts of organic matter in it not only holds more carbon, it also sequesters more carbon. The organic matter itself is carbon, so the more there is in the soil, the more carbon we are storing away. But there is a relationship whereby the plants growing in soils higher in carbon are actually able to draw larger amounts of carbon from the atmosphere, which is then exuded from their roots to be stored in the soil. This creates a positive feedback loop in which the soil is being continually improved by the plants, in turn supporting better plant growth, all the while sucking more and more carbon from our atmosphere.

Ok, so our little suburban garden will not stop climate change, but it is actually making a real difference. The more of our soil we have under healthy plant cover, the more difference we are making. For the soil to be the hero in reversing climate change, we need agriculture to get on board big time. Regenerative agriculture is finding favour all over the world and producing amazing results. Deserts are being turned into productive and fertile farmland without the need for fertilisers. Adding organic matter to the soil is the key to this process, although cessation of tilling and never leaving soil bare is also important. These same principles apply to our little home plots of dirt too.

The most exciting part of all this is that the simple process of adding organic matter to the soil, even if that is only through increased cover of plants, is working to improve the soil, and is also working to remove carbon from our atmosphere. Win, win, win! A study found that a one-off addition of compost to soil could cause a continual increase in that soil's carbon content. (https://www.marincarbonproject.org/). This works by kickstarting good growing conditions and letting the plants do the work.

Bare soil is a big problem. Bare soil dries out, the microbial life in it dies, it erodes badly and worst of all, it releases carbon into the atmosphere. Where there are vast areas of bare soil the carbon loss can be significant, and there is now plenty of evidence that the carbon release via agriculture rivals that of fossil fuel use. The more we leave our soil bare, be that in our garden or large-scale agriculture, the more we are destroying the soil and the climate. Even a covering of mulch will stop the damage to the soil and will stop the release of carbon to the atmosphere. The key to solving climate change could lie in a solution as simple as to cover bare soil with more plants. A world with more plants in it sounds fabulous to me and to every other gardener out there I'm sure.

All that sounds so easy doesn't it? So there is a good chance it is too good to be true! There are some scientists who claim this process will take hundreds of years and so will not save us, and there are scientists who claim this can be done within 30 years. Soil science is moving forward in leaps and bounds, so I guess we will know one way or the other soon enough. The one thing all soil scientists agree on is that caring for our soil is absolutely critical in maintaining a healthy planet. Soil that is bare and disturbed releases carbon dioxide into the atmosphere, so even if we cannot rely on soil to reduce the carbon in the atmosphere, we can at least treat soil with care and respect and in doing so limit the role of soil degradation in increasing atmospheric carbon. As gardeners we may only have a small patch of soil to care for, but since caring for our soil is also absolutely critical for good gardening, it really should be something we are doing anyway.

The role of mulch in soil care

As mentioned, bare soil is to be avoided at all costs. If we do not have good plant cover, most gardeners already know that we need mulch. Laying mulch is an important part in protecting and caring for soil. It acts as a protective blanket over the soil, helping to shade it from the hot sun, reducing moisture loss through evaporation and keeping the soil warmer in winter. A well-mulched soil can be up to 10 degrees cooler at 30cm deep than bare soil – a big difference for the plant's roots! Mulch also helps to protect soil from wind and water erosion. Bare soil will always be more prone to erosion, hold less moisture, contain less soil biota and will experience greater temperature fluctuations than mulched soil. Bare soil to me seems like bare skin burning in the sun – not a good idea!

Mulch plays a role in suppressing weeds. This is because many seeds need exposure to light to aid germination, hence there is always a flush of weeds in newly turned soil and another reason not to turn the soil if you don't need to. If you wish to grow annuals and have them self-seed, you may need to consider sowing seeds in trays and then planting them out into the garden, or else hold off on mulching until the seedlings are big enough to mulch around. A very thin sprinkling of a fine type of mulch such as dried loose grass clippings can work for seeds. Make sure when you sprinkle the fine mulch that it does not completely cover the soil. The tiny amount of shade and protection this provides can help seeds germinate by reducing drying out and compaction of the soil just enough to give the seeds a head start. Gravel paths are a great place for seeds to germinate – both valued garden plants that can be transplanted and weeds that are unwanted.

A well-mulched garden can create homes for little critters in the mulch layer and can give a neat, finished look to a garden.

Choosing mulch

Not all mulches are created equal. Organic mulches will break down and add organic matter to the soil. Inorganic mulches such as pebbles will not. There is a time and place for different mulches so choosing which one is best will depend on your situation. Of course, it does not hurt us to think about the environmental cost of the different mulches either.

Straw-based mulch

Firstly, let's look at the straw-based mulches. Sugar cane mulch is the most readily available here in Australia and is cheap and easy to use. It comes in easy-to-handle bales that most gardeners can manage to break up and spread easily. It is a by-product of the sugar industry, so turning it into a garden product added value to the sugar industry by creating a market for a waste product.

Good old-fashioned straw is also an agricultural by-product. It is usually the leftover bits when grain is harvested. Unlike sugar cane mulch, which is wrapped in plastic, straw is usually sold by the bale and held together with baling twine, which can be readily reused in the garden as plant ties. Any straw-like material that comes in a bale rather than in shrink-wrapped plastic will be coarser because anything too finely chopped would not hold together in a bale. I prefer to buy bales when I can to avoid the plastic wrapping. As this is messier, it is not as readily available, and most hardware chains and garden centres will not have it. You can find it at produce or farm supplies stores. It is often used for animal bedding. If you can get it when the animals are finished with it, all the better as it is now nutrient enriched. You will need to know the farmer for post animal used straw as it is not likely to be packaged and sold in this state!

Pea straw and lucerne hay are also great in the garden but are more expensive and for good reason. They are primarily grown as stock feed, and as such they are not a by-product but THE product. In times of drought or when stock feed is in short supply, the price will go up and it is unlikely to be available to gardeners. Both of these are legumes, so produce a protein rich material that is great fodder for animals – as well as your worms and soil.

A warning however with straw, pea straw and lucerne – they all carry the risk of containing weed seeds! Some of these can benefit your garden, but occasionally it leads to serious weed spread, so monitor your garden to see what comes up when using these products. Don't be too hasty in weeding however, it is quite common for lucerne to be the weed sprouting, and that is something you probably want to keep. If you let the lucerne grow you will not only have a very pretty garden plant with purple flowers, you will be growing a living, soil improving mulch.

These mulches are all very good for the soil and make the worms happy, but those such as sugar cane mulch, which are making use of a waste product, are the more sustainable. Be aware that if organic is important to you, you need to look for an organically grown product to avoid fertiliser or pesticide residue. Organic products are sourced from organic farms where the crop has been grown organically. As with everything organic – it is better to use less pesticides definitely, but to use less oil to ship the product is equally important, so if you have an option to buy a locally grown product even better, especially if it then comes without the plastic wrapping.

To be even more sustainable, grow your own mulch. Cut grass is what straw actually is, so if you are slashing a paddock, there is your mulch. The proviso here is that straw is long pieces of grass, not the small chopped up bits from your mower. I have taken big bags and collected up the huge piles of grass clippings the council leave behind when slashing roadsides and parks in summer and mulched my entire garden for free this way. Council slashers cut the grass more coarsely than do domestic mowers, so it can be used as mulch when it is dried and loose. Your backyard lawn clippings are too fine to use as mulch, they must be composted before adding to the garden or they will form a water repellent crust. You can also sprinkle alfalfa seeds around the garden. They make pretty garden plants, and fix nitrogen so they are good for the garden living or dead (and chopped and dropped). Alfalfa is what we call the sprouting seeds and as the plant grows, we call it lucerne – same plant, great mulch.

All of these straw-based mulches have a tendency to compact on the surface of the soil, restricting water penetration. Every month or so, or after heavy rain, get out with the fork or a metal rake and loosen them up again. The finer the material, the more likely they are to compact into a crust that repels water. This is the main reason that grass clippings need to be composted rather than used as mulch. The coarser product that you will get in bales will be less likely to form a crust than will the finer product you buy in plastic wrapping.

This group of mulches will break down reasonably quickly (unless they have compacted and are repelling water, in which case they won't break down at all), and this is actually part of the job we want them to do. They are adding valuable organic matter into the soil as they break down, thereby giving gradual but continual soil improvement. They do need to be topped up regularly.

Generally straw-based mulches are used in things like vegetable gardens or places with a lot of annual plants. This is because it breaks down quickly and can be replaced as the next crop goes in. It is rarely considered aesthetically pleasing but is treated as practical and used where aesthetics is not important, or at least can be compromised temporarily such as when planting flower seedlings. There is no reason why you cannot use it in all parts of the garden, although amongst more permanent plantings we tend to look for a mulch that we do not need to replace as often. For that we turn to wood chip mulch.

Woodchips

Woodchip mulch is the next most widely used type of mulch. It is an organic product in that it comes from trees that were once living, but it is not always a good choice for the environment. In most cases the bark mulch is made using just that – bark; a by-product of the forestry industry, which is not always as strongly regulated as it could be. In most cases bark is sourced in Australia but this is not always the case. Some of the exotic species are not grown everywhere and it isn't always the case that the woodchip or bark chip is a by-product or grown on a plantation. Ask your supplier if their woodchip mulch is FSC certified. This stands for Forestry Stewardship Council. Their logo appears on products derived from timber (including paper) to indicate that it is sourced responsibly.

In particular, beware of cypress pine woodchip, which is promoted as being termite resistant. In the USA, native cypress forests are being devastated for timber and woodchip, and much of the cypress mulch there is now produced from immature trees that have not fully developed the chemical components required to repel termites.

A drawback with woodchip mulches is that they occasionally attract termites, especially if they are applied too thickly or are left in an undisturbed heap. Play it safe and keep woodchip scraped back from the edge of your home. If in doubt, call an expert out to check your set up, or let the chooks free range near your house – they will eat any termites that try to come close. Termites hate to be disturbed, so if you are the sort of gardener who is always planting something and busy in the garden, you are less likely to have an issue. A thin layer of mulch is usually too easily disturbed to be attractive to termites, so keep your mulch less than 3cm thick if you are worried about termites. Keep in mind that in Australia we have over 200 species of termites but only a handful of these cause damage to homes. Termites play an important role in breaking down old timber – including woodchip, and returning the carbon to the soil. So long as you are sensible about keeping your house protected, they can actually be quite helpful little recyclers in the garden.

Another issue with cypress is that it is frequently dyed. The dyes used are most often non-toxic, however it is another process using energy, chemicals and water and is completely unnecessary yet adds significantly to the eco footprint of your garden.

Slash pine and hoop pine bark mulches are readily available from plantation trees. Used in excess slash pine can lower the soil pH, so I recommend making sure you always add some compost before reapplying if you are using repeated top ups of pine bark.

Another variety of readily available mulch is chopped trees directly from the tree loppers. Most landscaping yards have a version of this. The mulch is usually pretty fresh, and rather unpredictable. It will often contain large chunks as well as chopped up leaves. There is some discussion of it as a means of spreading weeds and disease, however I have never had any problems at all with it when used sensibly – that is, not too thickly especially around trees and timber structures, but this applies to all types of mulch. Healthy plants are far less susceptible to pest and diseases regardless, and I have yet to see any evidence that woodchip mulch from tree loppers carries or transfers disease to healthy plants and trees. In wet conditions it is common to see a variety of fungi in this type of mulch, but this is beneficial not harmful as it aids the breakdown process. I like the fact that there is some diversity to this type of mulch. To me it looks more natural, more 'forest floor'. Having a variety of wood and leaf contributes more to the soil in the long run – diversity of physical and chemical attributes means inconsistent breakdown, providing various homes for microbes and critters for longer, and a variety of nutrients added over time.

A big plus with this irregular forest mulch is that again it is a by-product. Sadly, those trees would be cut down and chopped up regardless of whether the mulch was to be used or not, so better to use what is available. It is also most often sourced locally, which is a big tick.

All bark mulches can cause nitrogen drawdown, but the fresher the mulch, the bigger the nitrogen drawdown will be. As woodchips start to break down, they require nitrogen to feed this process. The nitrogen is drawn from the soil and can leave it depleted, which can lead to garden plants becoming nitrogen depleted. You will see this in the form of yellowing leaves on the plants where you have recently added mulch. It is easily overcome by putting down a nitrogenous fertiliser such as a pelletised chicken manure product, or better still homemade compost, under the woodchip mulch. Fertiliser is generally best added under the mulch anyway, so this is good practice gardening. Much is said about how some mulches cause more problems than others with regard to nitrogen drawdown. In fact, all bark mulches will cause some degree of nitrogen drawdown as they start to break down, and all bark mulches will return that nitrogen and more to the soil once the decomposition process is underway. The degree of nitrogen drawdown is usually very small and short term. It is rare for this to cause any significant trouble for the plants in the garden, even if you do forget to add some fertiliser underneath. If you do see yellowing leaves on newly mulched plants, a watering can of compost tea will fix the problem. Place some compost into a stocking or old pillowcase and then soak it in a bucket of water until the water turns a rich tea colour. This is a great liquid feed for all sorts of scenarios but will provide a quick nitrogen fix if needed.

All of these wood-based products will add organic matter to the soil as they break down. They will all also leach some carbon back into the atmosphere as they break down. A happily growing garden will take up most of this carbon from the atmosphere, offsetting what the mulch releases, and the carbon which is released back into the soil can be stored there indefinitely. The greenhouse gasses produced when these timber waste products go to landfill instead are much worse, and we should see that while some carbon is released to the atmosphere, we are storing more of it in the soil, so achieving an overall good.

> It should be noted here that there are many natural processes that release carbon into the atmosphere. Most of these are not contributing significantly to atmospheric carbon levels, and yet they pop up as yet another something we should avoid – like using woodchip on the garden because it releases carbon into the atmosphere. Let's get a bit real. EVERYTHING has some impact. Having no impact at all on the earth means not existing at all. We are aiming here to do more good than harm in our gardening, not to delete our presence from the Earth!

Other organic mulches

There are other organic mulches that are sometimes used, although these are often more soil conditioners that have been used as mulch – mushroom compost, animal manures, grass clippings, fallen leaves, seaweed. These are all great for the soil but will be best if composted first and then covered with mulch. Leaf mulch is often very effective but can break down quickly, unless you have a lot of dry eucalypt leaves, in which case put them through a mulcher if you can and add them to grass clippings or animal manures to assist the breakdown of these very course leaves. Grass clippings should never be used as mulch until they have broken down or dried out, as the composting process will cause rot if close to the stems, and the fine particles will compact, forming a crust that stops water getting through.

Any mulch which contains many fine particles, including grass clippings and many commercially available wood chip mulches in the form of 'fines', are likely to become a problem. All fine mulches will crust and exclude water, making the soil hydrophobic and do more harm than good. If you find yourself with a fine mulch, you may need to go over it with a fork every so often to loosen it up. Sugar cane mulch falls into this group. It often breaks down fast enough to not be as much of a problem but keep an eye on it and loosen it periodically. A good mulch has large particles and an open nature which will allow for easy water penetration.

Inorganic mulches

There are also inorganic mulches that can be used – gravel, sand and shredded rubber. Shredded rubber is a recycled product being made from old car tyres, but contains many pollutants and compacts, so while it is great as soft fall in playgrounds, it is not great in the garden. You may find that shredded rubber works on pathways instead of concrete or pavers. It is not currently widely available for home use but is being used more and more in commercial and public plantings.

Sand is not all that commonly used any more, although I have heard of Italian old timers still preferring sand as mulch. The sand simply makes a layer on top of the soil that acts like a blanket, helping cool the soil underneath and retain moisture. If the soil is heavy clay, over time it will also help break up that clay. Sand and gravel are good options for areas where good airflow is essential – such as growing Mediterranean or arid plants in humid climates, as the mulch is reflective and will not trap humidity in the way that organic mulches will. These mulches will work to keep the soil cooler, but their reflective nature can make the plant itself hotter, which not all plants will appreciate, so they are best avoided for plants with large green leaves in a sunny position. In a shady position, the light colour of the mulch (e.g. sand or white gravel, or even sugar cane mulch) can increase the ambient light, making the space brighter and improving plant growth.

Sand and gravel are mined products. Mining sand on offshore islands in Queensland has caused much controversy, but in reality, sand and gravel are usually mined locally, and the vast bulk are used in the construction industry, not in landscaping. An exception of course is the highly decorative gravels which are far more expensive and have travelled much further to reach your garden. The bags they are sold in will usually show their country of origin. There may be issues regarding lack of environmental controls in those countries and there are also the travel miles to consider. If you don't really need to use those products, they are best avoided for purposes of sustainability.

Gravel does not contribute organic matter to the soil, so has no role in soil improvement. It also won't need to be topped up as it does not break down. Over time it may wash away or become pressed into the soil, but in general it won't need replacing like an organic mulch. If you have poor soil, it will be essential to improve it before mulching with gravel. Gravel is best used in situations in which it will be long term - a garden that is not going to be regularly replanted or need regular soil improvement. In this situation it is quite sustainable, because although it has a higher input of energy to dig it up, size and deliver it, that is all offset by its longevity in the garden.

Gravel mulches are recommended for gardens near homes in high fire risk areas. Organic mulches can become fuel for fires in extreme conditions. In this situation the mulch is too dry to break down and contribute organic matter to the garden anyway. Gravels which are made from crushed and recycled concrete and old bricks are also good as although there is an energy input into making them, they are not a virgin product, and are instead a waste product which is given a new lease on life instead of being dumped in landfill. These are often not considered as decorative and usually used as drainage gravel rather than mulch. Just because someone else thinks they are not sufficiently decorative to use above ground, does not mean they are not perfect in your garden. These are the sort of things that become a very personal choice.

Applying mulches

The way we use mulch can also impact on how sustainable our gardening practices are:

First – choose the right mulch for your application, as the wrong mulch will simply cause trouble and be a waste of resources. Generally, straw style mulches are best for vegetables, annuals and smaller gardens with new plants establishing because it is easy to use carefully around tiny plants and can be updated every season. Woodchip mulches are best for large garden beds, shrubberies, and any gardens which are already established, or which will have larger plants planted into it. Gravels should be used for xeriscaped gardens (those designed for extremely low water availability), or around the house in areas with high termite or fire danger, or for paths.

Second – set up the garden ready for the mulch. Remove weeds. To avoid using poisons, hand pull or cut very low and cover with damp paper before mulching. Add compost if you are using it, and nitrogenous fertiliser of you are using a bark mulch, then cover with mulch.

If you don't have many weeds to worry about, don't add the paper, as it will create a layer between the mulch and the soil that slows water penetration and oxygen transfer. Avoid weed matting for the same reason. Weed mat is not good for the soil – it takes a very long time to break down (if at all), and limits oxygen transfer with the soil, so the soil beneath can become quite sour. Over time soil accumulates on top of the weed mat and weeds find their way through it, so it does not offer such significant long term weed protection to warrant the unnecessary use of resources (plastics or geotextiles) and causes difficulties when trying to dig a hole and add a new plant to a garden later. Weeds such as nut grass laugh at weed mat, they will simply grow straight through it.

Third – water the ground well before mulching, even better mulch after a good downpour of rain. Getting water through the mulch and into dry soil can be a bit hit and miss, so much easier to have the soil nicely wet before you mulch so you know there is moisture in there worth protecting. Now is the time to fertilise as well, so that the fertiliser is in contact with the moist soil for easy penetration. Fertiliser applied over the top of mulch is largely wasted.

Fourth – not too thick. Five centimetres is a good depth to work to regardless of the type of mulch you are using. The thicker the mulch, the harder it will be for water to penetrate. Super thick mulch will be more effective at keeping weeds down, but it will also keep your plants down, so unless the garden is in a fallow period to be ready for planting next year, it is simply wasteful to put mulch on too thickly.

Putting it into practice

Now that we know how to care for our soil, how can we give this care as sustainably as possible? Keeping your source of organic matter as close to home as possible and wrapped in as little plastic as possible are the best ways to start. Chop and drop gardening, composting your own green waste and kitchen scraps, using your pets' recycled paper litter after they have added some extra 'flavour', raking your own leaves (or your neighbours') are all easy, free and sustainable options.

Local mowing contractors will often very happily drop the day's grass clippings for free if you a have a convenient place for them to dump it. By dropping it locally they save the time (and petrol) to dispose of it at the council waste facility, and you get a free local resource. If you are concerned about organic gardening this may not be a good option for you as it is likely the grass has been treated with inorganic fertilisers, herbicides and pesticides. These are usually not present in large enough quantities to cause any issues in your soil, but in strict terms will mean you would not get organic certification.

This actually applies to all animal manures as well. Unless they are sourced from an organic farm, they may contain traces of antibiotics or other pest treatments that the animals have been treated with. Usually not in high enough amounts to have any impact at all in general garden use, but not strictly organic all the same.

I have heard some reports that treatments for intestinal worms and other insect parasites can come through in the manure and kill earthworms and other soil microorganisms. I use a huge amount of aged stable manure and so far, have never seen anything but a dramatic increase in earthworms and soil biota for it. A lot of these products are broken down in the gut and then in the aging process, so if you are worried, always compost the manure before use.

Bagged animal manures are often sold cheaply at roadside stalls and often provide the family's kids with pocket money by bagging it up, so if you are driving that way anyway, grabbing a few bags is highly sustainable (plus the bags themselves are usually reused feed bags). Buying a bagged product from a hardware or nursery involves a commercial composting process, transport and packaging, but is still better than having this waste material sent to landfill, dumped in bushland or allowed to leach into waterways.

Garden Products for the Soil

The shelves of hardware stores and garden centres are full of them, garden experts tell us we need them, and somehow, we end up with a shed full of them. Many gardeners don't really understand what many of these garden products are or when to use which product. This is an area of mass choice and consumerism in overdrive. I know plenty of gardeners who use the 'when in doubt throw it all in' principle. Given how problematic some of these products can be, that is a scary idea. We tend to think fertilisers are good for plants, but too much can kill plants pretty quickly. If nothing else, the volume of packaging from all these different products should be an indicator that something is out of whack.

We have all seen some pretty spectacular forests without a tub of fertiliser or bag of gypsum in sight. Nature can create incredible growth without fertilisers, but to be fair, our garden settings are a little different. The main difference is our expectation. We expect to grow what we like, not what suits the location. We want it to grow fabulously, and we want it to grow quickly.

Later on we will look at how we can garden more naturally with great success, but for now, let's have a closer look at what some of these products are and what they do. If we know a little more about them, we are better positioned to make an informed choice about what to buy and how and when to use it. This approach will not only cut down on excess consumerism and unnecessary chemical use, it will also save us time and money.

Soil additives

Any product you add to your soil is a soil additive. Our final aim with all additives is to make plants grow better. As we have already established, great soil makes plants grow great, so as gardeners, we tend to add lots of things to the soil with the hope it will be the magic ingredient that makes our plants flourish - fertilisers, clay breakers, wetting agents, trace elements, lime, dolomite, cocopeat, activated charcoal, the list goes on.

Let's have a look at some of the commonly used products for soils and potting mixes and provide some thoughts on their sustainability credentials.

All mineral based additives and fertilisers are mined or are manufactured from mined minerals. Dolomite, phosphorus, magnesium (including Epsom salts which are commonly used for yellow leaves), iron chelates, sulphates, nitrates, gypsum, lime, rock dust or rock minerals – all are mined products. Even products like vermiculite are natural minerals which are extracted from the earth and then superheated to form the product we use.

All of these products have been extracted from the earth, refined in some way through energy intense processes, packaged and shipped. All have an environmental footprint to some extent. The easiest way to reduce this footprint is to simply not use them. A lot of the time we don't actually need them if we are caring for our soil well. If we are going to use them, here are some guidelines to help with your decision as to which ones to spend your money on:

- buy local as much as possible to reduce transport miles.
- buy the most efficient bag size you can – a larger bag is usually better value for money and less packaging but is wasteful if you only need a little.
- be sure of what you want to achieve with the product and choose the best product for that outcome. If in doubt, use compost instead.
- choose a single product that performs many functions over numerous individual products where you can.

With this last point comes the reason we are using these additives. One of the common reasons is to fix a single nutrient deficiency, in which case we go to the hardware and buy a single product, for example, iron chelates. A few months later we identify a different deficiency and go and buy another. Much of this would have been reduced if in the first instance we purchased a trace element mix and treated a variety of issues with one product.

Fertiliser

Fertiliser use is a key area we can make sustainable choices. Simply by better understanding the role of fertilisers we can garden more sustainably. In general, fertilisers are massively overused. This overuse can create a snowballing of other garden problems which have us running to the stores to buy more products to fix each subsequent problem. This of course leads to an escalation of garden related consumerism and a cluttering up of the garden shed with half used products rapidly going out of date.

In essence a fertiliser is any product that supplies plants with the nutrients required for growth. This usually consists of the three major plant nutrients – Nitrogen (N) which is required for healthy leaf growth and as leaves are usually the most voluminous part of the plant, is usually the element plants need most of; Phosphorous (P) which is required for the development of roots and flowers; and Potassium (K) which is required for providing plant strength and disease resistance. Together these are known as the N:P:K ratio and are listed on every fertiliser packet.

Some of you may be wondering why I have not said that potassium is needed for flowers, after all, it is potassium in the form of potash that we put on plants to increase flowering is it not? Well actually no, it is phosphorous which promotes the development of flower buds, which is why farmers have loved their superphosphate over the years. Potassium creates stronger healthier plants, and of course healthier plants will flower better. Potassium has also been much less demonised than phosphorus as it is less environmentally harmful. Phosphorus is often applied in a form which is not readily plant available and leaches out of the soil into waterways where it has contributed to blue green algal blooms.

Another issue with phosphorous is the sensitivity of many Australian natives to high levels of it. Some Australian soils are naturally very deficient in phosphorus and some native plants have developed highly sensitive root systems to extract tiny amounts of this nutrient from the soil. Adding more phosphorus to these plants is like shouting into a megaphone right next to someone's ear – just far too much. Not all Australian soils are phosphorus deficient however, which is a surprise to many. If your local native vegetation does not feature plants in the Proteaceae family (grevilleas, banksias and their relatives), your soil is not particularly phosphorus deficient, and your local natives will be less sensitive to phosphorus in fertilisers.

On this note – have you ever looked closely at the N:P:K ratios on fertiliser tubs and wondered why some of the fertilisers for natives have more phosphorus than do some of the non-native fertilisers? That should be your first indication that many of those specialist fertilisers are a con.

Nitrogen is the first number in the ratio and is usually the largest number of the three. As mentioned, plants do need more nitrogen than other nutrients, so this makes sense. However, plants do not need as much as is usually given. Nitrogen is highly water-soluble and is easily leached out of the soil by rain or over watering. Overuse of fertiliser, especially on large-scale agricultural settings, has been shown to release nitrous oxide, a very powerful greenhouse gas. When plants are unable to take up the large amounts of nitrogen applied through fertilisers, that nitrogen becomes a pollutant in waterways and in the atmosphere.

High nitrogen will cause excessive plant growth – in the garden, or wherever it has leached to, which is often a nearby waterway. Nitrogen is a major water pollutant, so care should always be taken when using fertilisers in any garden where runoff can reach a nearby waterway.

Nitrogen is also one of the components of fertiliser that causes it to burn plants if too much is used. Any product with greater than around 5% nitrogen could cause plants to burn when applied generously. Many lawn fertilisers contain more than 20% nitrogen, which is how overzealous gardeners can accidentally over-fertilise and damage the lawn they are trying so hard to care for. Like lawns, plants also like nitrogen but will perform much better (and with less risk of damage) if small amounts are applied often rather than large doses occasionally.

In fact, nitrogen is the element we really overdo badly. Plants grown with an excess of nitrogen will grow quickly and be lush and green, which to many is the desired outcome. The downside is that without the balance of other elements, plants put on soft, sweet growth, which is weak and highly attractive to pests. Here begins the snowballing of other problems in the garden. Nitrogen to plants is like sugar to children, in small doses it gives great results, but overdo it and trouble is coming.

Another issue with using fertilisers, especially inorganic fertilisers, is that they are often high in salts. Overuse can (and does) lead to salinity issues in soils.

All chemical fertilisers are based on fossil fuels and the mining of minerals. Organic fertilisers are usually based on by-products (manures or spent mushroom compost) from farming. This immediately creates a point of difference from a sustainability context. How do you know which is which? Any fertiliser in the form of a little ball or granule (usually white but often coloured) is inorganic. It is chemically synthesised. An organic fertiliser will be in the form of a brown pellet and will usually have some smell associated with it.

Are you one of those gardeners who use more fertiliser than you really need? Halve your fertiliser use for a whole year and see what happens. Notice if your plants suffer for it. Try to observe any reduction in sap sucking insects because of it. Maybe the only difference you will notice is a little less pressure on your gardening budget. Feel free to write and let me know how you go!

When it comes to fertiliser, it is not only how much we use, but which one we use. This is a great example of companies exploiting consumerism for their own profits. Have you ever bought different fertilisers for different plants? One for the roses, the fruit trees, the lawn, the vegetables, the natives and a general purpose one for the rest of the garden? Go to the shed and read the details on the back of the labels. I am very sorry if you now feel conned, but I hope that I have saved you when you next go shopping for fertiliser. I have spent the time in stores evaluating all the different packages when writing articles on fertilisers. I was shocked to find that some companies don't even try to make their various products different. In fact one company had five differently labelled products, three of which had identical nutrient make ups, and the other two varied only very slightly. This is simply an exercise in exploitation to my mind. But perhaps it is more symbolic of the direction we have let the industry take us.

As gardeners we know enough to know that different plants have different requirements, but in general we know very little about what those requirements are. As I have said before, this is as much a consumer industry as any other. Fertiliser companies know that most of their customers will read the front but not the back labels, and that even if they do read the back, many will simply not understand it. We do not need to be experts, but we can certainly save ourselves money and reduce environmental impacts by choosing one good general fertiliser and then using it sparingly. Which begs the question – what makes a good fertiliser?

As I have mentioned, a fertiliser with 5% or less N is a good place to start. Once the total N is over 10% you really should put those products back on the shelf. A good fertiliser will have small amounts of a lot of things, not just the N:P:K. Often this starts with sulfur (S) but may also include calcium (Ca), magnesium (Mg), iron (Fe), zinc (Zn), boron (B), and other trace elements. Plants need a good range of trace elements to perform well, not just the three major nutrients. A fertiliser that contains a mixture of mineral elements will reduce the need for supplementary addition of trace element mixtures or for treating mineral deficiencies.

A fertiliser that includes some organic matter will help feed the soil, not just the plants. The organic forms of the nutrients are both plant available, less toxic and less likely to be quickly leached. Organic fertilisers are naturally slow-release. These organic fertilisers usually also contain beneficial soil microbes which further support the health of the soil.

Inorganic fertilisers give nothing to the health of the soil and can often be detrimental as the high salts destroy soil life. Healthy plants require healthy soil, and fertilising plants without consideration of the soil is simply not holistic or sustainable.

There is also a plethora of liquid fertilisers available. If the product has an N:P:K listed on the front of the package, it is a fertiliser. Liquid fertilisers are designed to give gardens a quick boost, rather than a decent feed. They should always be used carefully and sparingly, and never make them up too strong. An experiment I did with my son's science class many years ago was a real eye opener for everyone involved. The children were testing different pollutants on plants they had grown, and were pouring all sorts of substances (good and bad) on to their pot plants. Over the next week we monitored the impact this had on the plants. The first plant to die, and it did so within 24 hours, was the one the kids had put liquid fertiliser on. Even the plant they poured petrol on lasted three days. Of course, they had made it up too strong, but plenty of gardeners do the same.

I personally do not use any commercial fertilisers at all in my own garden. Where plants look like they need a top up of nitrogen, they get a bucketful of compost from the chook pen (NOT straight chook poo). I recycle green waste through chop and drop gardening as much as possible, which keeps nutrients in constant cycle through the plants and the soil. Occasionally I give everything a good dose of a high-quality, locally made rock mineral product. My preferred rock mineral product contains no nitrogen at all, but it does include nitrogen fixing bacteria, plus 60 different minerals (including silica) so I know it is doing my soil the world of good, and in so doing is also working wonders for my plants. As I do not have a large output of crops, prunings or lawn clippings leaving my garden, I have much less need to replace the nutrients lost this way. This means the trace elements I add with the rock minerals are being recycled in the garden and there is less need to constantly top them up.

By recycling my own green waste and adding a bit of homemade compost (which includes chook poo) I am maintaining an ongoing, small but sufficient, supply of nitrogen to my garden. This is helped by the nitrogen-fixing bacteria in the soil, planting legumes throughout the garden to fix nitrogen into the soil, and of course the little bit of nitrogen washed out of the atmosphere by rain. Other products found in the fertiliser aisle are actually soil conditioners, plant tonics or soil additives. If it does not have an N:P:K ratio, it is not a fertiliser, but that does not mean it is of any more or less value in the garden.

Blood and bone

For a long time, blood and bone has been seen as a great source of nitrogen and phosphorus for the garden. It used to be a by-product from abattoirs, and when it was, it was worth using. Very few gardeners are aware that this is no longer the case. It is no longer dried and ground up waste from meat processing. That waste is now sent overseas for processing and extraction of anything of value. What is left is basically just carbon, sawdust almost, with no goodness. To create a garden-worthy product, blood and bone now has nutrients added back into it. It is no longer a natural by-product and instead is heavily modified and manufactured. If you look closely at the packaging it is also now labelled as 'Blood and Bone based fertiliser', and it has an N:P:K ratio listed on the packet. Quite apart from the high degree of processing required to make this product, and huge transport miles to ship it overseas and back, if you read the nutrient break down you will see there is no real point to it. It offers so much less than any other fertiliser available and offers nothing they do not. It is still popular because of what it once was.

Plant tonics

The commonly used seaweed and fish emulsion products are plant tonics. They contain trace elements, enzymes and microbes, which support healthy soil and healthy plant growth. Think of them a bit like a multivitamin for plants. They can be very useful in supporting plants that are stressed. This includes reducing transplant shock, or helping plants that have been blasted by extremes of weather. If you are trying to rescue a sad or dying plant, a plant tonic is the place to start.

In general, foliar plant tonics such as seaweed solutions are very effective in improving general plant health, and for stimulating root growth on young plants. Homemade compost tea, weed tea and worm juice are also effective, and a whole lot cheaper. I have read reports and discussions claiming that one is better than another, and every argument has something to contribute. In the long run, they are all beneficial. Homemade versions can be enhanced by adding a good mix of different weeds, or even seaweed collected from the beach if it is available (and if local regulations allow it). Some gardeners swear by using only comfrey or nettle to make weed tea, and these are both excellent alone or in combination.

Any plant with a high mineral content (which does include most weeds) can be soaked in water for a few weeks until the water is foul and smelly. At this stage it has not only broken down in the water, releasing nutrients from the plant material, it will also have fermented a bit, allowing for growth of microbes. Strained and diluted, it makes an excellent plant tonic. Sure, the garden will smell awful but only for a day. You can pour it straight onto the soil rather than over the leaves of the plants to reduce the smell. It is still doing a lot of good when added to the soil, as it is feeding and adding to the microbial life that is there.

If you add some nitrogen in the form of chook poo, or even fertiliser pellets to your brew, you will create a liquid plant tonic that is also a fertiliser. Don't add too much nitrogen – it can be beneficial, but a little goes a long way.

Most plant tonics are applied as a liquid. They are bought or made in a concentrated form and then mixed with water before being used. Because they are quite mild, they won't burn (unless you use too much nitrogen). This means you are less likely to do the harm you would with a liquid fertiliser should you get the measurements wrong. Mix it up until it is the colour of weak tea, then splash it around to your hearts content. By making your own, not only are you recycling waste products in situ, you are saving money and resources through processed and packaged products.

Like everything else there is a place for it, and particularly in a garden where plants are under stress it can really help, however there is little benefit to be gained by applying more than the plants actually need. On the flip side, if your plants actually need regular doses of such plant tonics, perhaps you should go back to your soil, and make some improvements at ground level.

Seaweed

Seaweed has been shown to have all trace elements required for plants, and it has also been shown that by foliar feeding, much smaller doses will be needed than by adding to the soil and feeding the roots.

There is some debate as to whether seaweed has trace elements in high enough doses to do the good that it is claimed to do, but it does also contain enzymes and vitamins which are highly beneficial and play a role in plant health through providing antibacterial benefits, speeding up photosynthesis, and feeding microbes in the soil which also benefit plant health.

Seaweed is a renewable resource, although it is under threat in certain areas and there are some doubts as to how sustainably it is harvested. The farming of seaweed has increased enormously, mainly for the food and agar industries, but this has a flow-on benefit for smaller users such as gardeners. The fact that such a valuable gardening resource comes from our oceans is yet another reason why we should all value the health of our oceans and what gets dumped into them. If you are buying a seaweed product shrink-wrapped in plastic film (a significant ocean pollutant) you need to stop and think again.

General sustainable consumer principles apply here in making smart purchasing decisions. A dried seaweed extract is available as an alternative. The powdered form means that transport and packaging inputs are reduced as the heavy water component has been removed and the product is worth considering from this point alone.

You can of course make your own with seaweed collected from the beach. Different states have different regulations about collecting seaweed, with most putting limits of around one or two shopping bags per person - so find out before you start collecting. It can be thrown straight onto the garden, added to the compost or added to compost tea. There is no need to wash it first as it is not high enough in salts to be a problem (unlike many commercial fertilisers!).

Soil conditioners

Soil conditioners are exactly what they say – products which change the condition of the soil. This includes clay breakers such as gypsum and products such as lime and dolomite that alter pH. Compost is a soil conditioner as it changes the structure of the soil. Rock minerals and biochar are also soil conditioners. Even wetting agents will fall into this category as they are changing the ability of the soil to hold water.

We are regularly told of the importance of using soil conditioners to improve our soil. If we have problematic soils, we may indeed need a bit of extra help. In most cases the sustainable soil care we discussed earlier will do the job of all of these products. Compost, whether homemade, packaged or naturally occurring, will add organic matter to the soil. That organic matter will improve the structure of the soil and buffer pH and other toxicity issues within the soil. To a certain extent it will also add minerals and trace elements to the soil. Use compost first and then decide if you really do need any further soil conditioning products.

Lime and dolomite

Garden lime is calcium carbonate. Dolomite is calcium magnesium carbonate. Do not confuse them with builder's lime, which is calcium hydroxide, is highly reactive and will burn plants and our skin.

Garden lime and dolomite are both used to raise the pH of acidic soil. They are often recommended for use in vegetable gardens where a pH of 6.5 – 7 is ideal. Both also add calcium, which is very valuable in a vegetable plot. There are of course other ways to get calcium, such as in egg shells, rock minerals or crushed shells, but these won't change the pH. In most cases, your garden will need the calcium more than the change to pH. If you do need to raise the pH, they are both very effective, although the dolomite is a little slower and gentler. These products are basically pulverised chalk or limestone. Lime will react with nitrogen, so if you are adding lime to your soils, wait a week or two before adding fertiliser.

Gypsum

Gypsum is also a form of calcium – calcium sulphate. It is primarily used for opening up heavy clay soils. It is only effective in some types of clay soil, so check which sort you have before rushing out to buy it. Put some of your clay soil in a jar of water and shake it until the water is cloudy. Let it sit for 10 minutes. If the water goes clear and the clay has settled, it will not respond to gypsum. If the water stays cloudy, gypsum should help. A light sprinkle of gypsum isn't enough. You will need about two cups per square metre. Dig that in and repeat the test. If the water goes clear in 10 minutes you won't need any more gypsum, but if not, you do.

Adding more than you need will actually cause harm to your soil and can make it set like concrete. In truth, most gardeners sling it around without thinking about how much they need or repeating the test to see if they have enough. More often than not, gardeners do not use enough to get the result they want. Those who have used too much often don't realise that it is the gypsum that has caused the harm then add even more thinking they have not yet added enough. Given the effectiveness of organic matter in opening up clay soils, and the added benefits of compost, perhaps you will be better off adding large amounts of organic matter to your heavy clay soils for good long term soil improvement.

Rock minerals

Rock minerals are touted as the organic solution to fixing trace element and mineral deficiencies in soils. They are sometimes touted as miracle fixers to all soil types. The quality and the ingredients of different rock mineral products vary enormously. While they are not a renewable resource, and have significant energy inputs in terms of mining, grinding up, packaging and transporting (sometimes internationally), they are often based on by-products of mining and quarrying.

A good rock mineral product can be highly beneficial and can replace the use of many other products – all of which are also resource intensive. An annual application in most gardens will certainly go a long way to balancing other deficiencies which you may otherwise be trying to diagnose in order to find the right additive (or two!). They can greatly improve overall soil and plant health and can replace the need for gypsum, lime, dolomite, wetting agents and a whole host of other soil supplements – IF they are high in silica. The one I use is based on silica, a mineral highly lacking in Australian soils, and one that is vital to plant strength. Silica plays an enormous role in regulating soil structure, making it valuable in both clay and sandy soils.

I find that by using this once a year, my need to use any other products at all (including pesticides and fertilisers) is almost eliminated. In this way, while the rock mineral product may not be completely sustainable itself, the greater outcome is achieved not just in improved plant health but in massively reduced need for many other products, greatly reducing my consumerism and saving me money.

There is some science that shows that soils which are high in minerals are better able to sequester and store carbon. There is that carbon storage idea again, suggesting that minerals may play a part not just in improved plant and soil health but also in the entire carbon cycle, the complexities of which we are still unravelling.

When it comes to trace elements, I once again prefer to use rock minerals, as this gives a far more comprehensive mix of trace elements than any manufactured alternative. A good rock mineral product will also contain a rich enhancement of soil microbes. Soil microbes are essential for turning the crushed rock into plant available nutrients, and no rock mineral product will be effective without them, unless you are supplementing them from compost or other sources.

I have found the rock mineral product that I use to be absolute magic in the garden. I personally use it in many gardens with highly variable soil types. I use it whenever I see a plant in need of care, be that fertilising, pest treatment or preparation for extreme weather. I sprinkle it over the plant instead of diatomaceous earth (which is silica and which acts to desiccate the sap sucking pests thereby killing them), I use it when I have weed problems or compacted and scalped lawns, I use it to deter pests and best of all I use it to make sure I am growing highly nutritious food.

There are two other huge benefits to using one product in place of many - cost and knowledge. There is no need to be an expert on garden problems if I have one magic cure-all. Now all those half full packages and bottles in the garden shed no longer need to take up space on the shelf, in your mind, or in your budget!

There is one thing lacking in using rock minerals – organic matter. All soils need to be topped up with organic matter. So even though they have replaced the need for almost all packaged products, I do still add organic matter to my garden via compost and organic mulches. Of course, as the plants takes up the additional minerals, all my chop and drop gardening is recycling those minerals back into my soil to be used all over again.

Bio char, activated charcoal

This is another product that is getting a lot of attention, especially in organic gardening circles. It is basically charcoal which has been activated with microbes. It can be very effective at creating porosity (air spaces) in soil, and because water and nutrients bind well to the charcoal, it can be very effective at holding moisture and nutrients in soil, but so can any organic matter. It also stores carbon in the soil, as does organic matter.

Basically, IT IS organic matter, being burnt wood. It works very well, but then so does organic matter, which has not been burnt. It is a great product but like others, it has significant energy inputs with added travel miles, and is easily substituted with homemade compost. However bio char has the advantage of being able to be dug into the soil for almost instant results. The charcoal breaks down slowly, so it does not need to be replenished as often as compost does. Compost does take time to alter the condition of the soil. Bio char can add porosity to the soil from the moment it is dug in.

Bio char can be particularly useful in sandy soils. Sandy soils tend to leach terribly, and struggle to hold onto nutrients and water. They tend to require huge amounts of organic matter on an ongoing basis. Bio char can be a very effective way of holding nutrients, microbes and water in the sandy soil, and in doing so can make those applications of compost far more effective. Biochar has also been found to be very effective in acidic soils which are often nutrient poor.

Bio char has been used agriculturally by traditional people around the world for centuries, but is only recently getting much attention in western agriculture. On a commercial scale it can be made by super heating waste organic matter in a low oxygen environment. It does not have to be made from wood; it can be green waste, animal manures and even sewerage sludge can be used to create biochar.

You can make your own. While I do not recommend setting up a charcoal manufacturing process in your garden, many indoor fireplaces and outdoor firepits do not completely consume the wood, leaving behind charcoal. This is more likely to happen if you do not have good air flow in your fire pit, as oxygen is required for complete combustion. To activate the charcoal, soak it in worm juice or homemade compost tea for a week or two. This will allow it to soak up both nutrients and microbes. By digging this into the garden you will be putting it to good use.

The ash from the fireplace is another matter. This has a high pH and if continually used in one place will alter the pH of your soil. Put it to use around those plants that like a high pH, otherwise add it to the compost.

Because the organic matter locked up in biochar breaks down very slowly in soil, it is being heavily researched as a means of storing carbon in soil. It has certainly been shown that by using biochar in combination with fertiliser, the nitrous oxide produced by the fertiliser is trapped in the soil and not released into the atmosphere where it is a serious greenhouse gas.

Fibre products

Let's look at some of the fibre products available. These include coir or cocopeat, peat moss and sphagnum moss. They are used in baskets, in special mixes for orchids and ferns or even to add organic matter and water holding capabilities to soil.

Peat moss is a fine fibrous product which is dark in colour. It is sourced from the dead moss accumulated underneath the living sphagnum peat bogs. Sphagnum moss is a spongy light fibrous material harvested from the living part of the sphagnum moss. Both living and dead cells of the sphagnum moss can hold 16 to 26 times its dry weight in water. The acidic nature of the moss limits fungal and bacterial growth, making it desirable in growing mediums.

Peat moss used to be a widely regarded product, until it was made known that the peat bogs of Europe where it was extracted were not regenerating. They are a very specialised natural environment where organic matter is laid down in cold, wet, anaerobic and acidic conditions which slows the decay process resulting in a light organic material which is capable of holding huge amounts of water. The key here is slow – formation of peat bogs is a very slow process. Not as slow as forming oil, but certainly much slower than growing a tree.

There is some debate around the world as to how peat bogs can be sustainably managed. In New Zealand they have sustainability management plans for their peat bogs, and all harvesting is done by hand using pitchforks with no heavy machinery allowed near the bog. In Canada there is evidence that some bogs are replenishing faster than they are being harvested, but others not so. It has been estimated that in Europe 90% of the bogs have been damaged or destroyed.

In Australia, sphagnum moss occurs in New South Wales, The Australian Capital Territory, Victoria and Tasmania at high altitudes and in isolated pockets. All sphagnum moss produced in Australia is wild harvested, but trials for growing the moss specifically for the horticultural industry are under way. It would be great to see these trials succeed.

There are discussions underway in the UK around a complete ban on peat moss products in the horticulture industry. Peat bogs there are badly degraded, and many species are now endangered as their peat bog habitat is destroyed. It has also been shown that peat bogs can store significantly more carbon than can forests, but when degraded they release carbon to the atmosphere. A voluntary campaign to reduce the demand for peat products has not been successful so it seems a complete ban on both locally sourced and imported peat products is on the cards for the UK. With that in mind, is peat moss really the best product for you to use in your garden?

Coir or cocopeat are a great alternative to peat moss or sphagnum moss, and for now at least, are far more sustainably sourced. Coir or cocopeat are by-products of the coconut industry, so once again we are seeing a waste product converted into a valuable new product. Coir is the dust and fine fibrous material in the husk of the coconut. Once the nut has been removed for the food industry, the husk is often utilised by the fibre industry for making ropes, brushes and matting. Even then, the fine particles are waste and can be used for the horticultural industry. Australia does not have much of a coconut fibre industry so the husks from our fledgling coconut industry are almost all used for coir for horticulture. As a crop, coconuts are a growing industry in northern Australia and it is being found that high yields can be had on marginal land and that it can be a quite low input crop in terms of fertilisers etc. Coir then is a very environmentally friendly product to use, although for now, most of it is produced internationally.

If you are lucky enough to have lots of old staghorns – how wonderful! If you have had to remove them and cut off the back material to re-hang them, make a point of keeping that old dead material. Chop it up and it makes a perfect homegrown alternative to either coir or sphagnum moss.

Vermiculite, perlite and pumice

These are all inorganic and inert products used extensively in horticulture to aerate soils and potting mixes and help hold water. They all have very good porosity – lots of little holes that provide air spaces in soil. These particles all also tend to hold water well around their outside, making it easily available to plants. They work similarly to charcoal with the difference that they are inorganic.

They are all slightly different, so to give a brief intro into them all:

Perlite is volcanic glass which is heated until it pops like popcorn, creating a very light product that looks like small balls of polystyrene. It has excellent porosity but can also hold significant plant available water around its outside. It is often used in seed raising mixes, and in garden beds to help aeration and to retain moisture.

Pumice is a graded stone, a very lightweight and porous volcanic rock formed during violent eruptions and is used mainly for the aeration of mixes. Pumice products should be certified as dust free as the fine dust will become concrete like when wet, having exactly the opposite of the desired effect.

Vermiculite is a hydrated magnesium aluminium silicate compound that is heated to make it expand and separate into layers which then allow it to hold water and provide aeration. Vermiculite has a particularly high water holding capacity.

These products are all considered safe to use in organic gardening as they are all naturally occurring products. They all have extensive use outside horticulture in various industries, and within horticulture they are used particularly in plant breeding and propagation nurseries, in aquaponics and in specialist mixes. I know of home gardeners who use them as a soil conditioner, or even as mulch. Their porous nature means that not only do they have excellent porosity and water holding capacity, they can also create good breeding sites in the soil for soil microbes which in turn increases soil health.

While there is no doubt that these products can have many benefits in the garden, from a sustainability point of view, are they really necessary or are we simply using resources excessively? These products are all mined, two of them have high energy inputs due to high heating of the raw product, then are bagged and transported around the world.

This does not make them bad, and by choosing not to use them, we will not really impact the total industry as the proportion of their total supply to the home gardener is very low, but sustainability is about treading a little more lightly on the earth where we can, and one way is to stop using products, even good ones, that we just don't need. For my money, locally sourced organic matter will achieve the same results with a lower ecological footprint.

> *Air in Soil*
>
> *So far, I haven't discussed the importance of air in soil to any great length. We know we need to water in new plants to ensure there are no air spaces around the roots which will dry them out, but even when the soil is watered, healthy soil contains very tiny air spaces, and this is known as the porosity of the soil. Plant roots actually need to breathe air, as well as absorb water and nutrients from the soil. Air spaces in the soil make oxygen available to microbes and keep the soil light enough for plant roots to push their way through.*
>
> *A major problem of both waterlogged and compacted soils is the lack of oxygen in the soil. My preferred method of aerating soils is to let the worms do it for you. This saves my back and also the natural soil structure. Badly compacted soils may need a helping hand to loosen them up, but otherwise add organic matter and compost and the worms will come and do what they are designed to do - aerate the soil.*

Water crystals

Water crystals are a synthetic polymer, a sort of plastic. They can be useful to store water in the soil (soak them in water BEFORE putting them in the soil), but they do not hold oxygen and can turn sour. Why use such an unnatural product when there are much more environmentally friendly alternatives available? Or better still, ensure your soil has sufficient organic matter so that none of these products are needed.

Add compost instead

Whichever soil additive you like to use, take a step back and look at your compost first. Can you achieve the same or similar results using your compost more effectively? Would an occasional dose of rock minerals make a difference to the quality of your compost? Do you need an occasional helping hand or a lot of help? Do let budget constrains influence your decisions. If you are spending a small fortune on various products, there is probably a better way. Homemade compost doesn't cost anything. Collecting leaves and other organic matter is either free or cheap and local, and is putting a local resource which is often seen as a waste product, to good use. Win-win! If there is any doubt about which product to use, add compost instead.

Potting mix

Go to any large hardware store or garden centre and you may be surprised at just how many different potting mixes are available. As a purchaser we will look at the price when making a decision and then look at what it is recommended to be used for. I have yet to find a cheap potting mix that has not cost me more by killing the plants I plant into it than what I saved on the bag of mix in the first place.

ALWAYS go for a high-quality potting mix. This is usually the most expensive one and this is the one you should use. You can often save money (and packaging) by buying a larger bag size. Cheap products that do not work are not just a waste of money, but a waste of energy and resources as well.

Cheap potting mixes often have a lot of large organic particles in them, usually bark. This makes them too free draining and lacking in anything substantial that the plants can root into. A good quality potting mix will be able to drain, but also have some water retention properties and will have a reasonably fine texture to allow the fine roots of potted plants to penetrate easily. Here we have Australian Standards, which are symbolised by a series of red ticks on the package. While this is an indicator of quality, it is not in itself a guarantee that the potting mix is the best you can get.

Another aspect of potting mix which influences our purchasing decision is the specified use for the potting mix. Various formulations exist for orchids, succulents, fruit, vegetables, flowers and rose and gardenias. The orchid mix usually contains more chunky bark for exceptional free drainage for orchids. The succulent mix often contains a lot of small bits of bark and some sand for free drainage. The rest should be formulated with fertilisers and pH to make them more suited to various plants.

This sounds so helpful, but is it really? Nope. A good quality potting mix will work very well for everything. It may need some bark added to work well for orchids, or some sand added for succulents (although I confess, I use a good quality potting mix as is for succulents and it works very well). I have potted roses, with a few vegies in the bottom of the same pot, all growing very happily in good quality potting mix. Don't be conned into buying more than you need.

All potting mixes will deplete over time and need topping up and refreshing. The easy option is to take the top layer of old mix out of the top and replace it with new mix. Rather than replacing the old mix, it is entirely possible to refresh the mix a few times before it needs replacing. Worm castings are gold for this if you have a worm farm, or well-rotted compost. Poke some holes into the old potting mix and then top up with the fine compost or worm castings and water it in gently. It will go down into the holes and work miracles. This is especially good for very large pots that are too big to easily re-pot. Every so often it is a good idea to water pot plants with a diluted solution of weed or compost tea, worm juice or a seaweed solution.

Pot plants do need some form of fertilising. Unlike garden plants that can send their roots searching for goodness, or deeper into the soil for moisture, plants in pots are limited to what you give them. This means pot plants will always need more care than plants placed in the ground. I use homemade compost to top up my pots with great success. The benefit of using compost is that it tops the soil level up in addition to providing nutrition. All potting mix will sink over time as the organic matter in it breaks down.

Don't use potting mix in the garden. It is designed for use in pots. It is too free draining to use in a garden bed and will quickly become dust dry and hydrophobic. If you need to improve your soil, refer back to the chapter on soil, but do not use potting mix. If you need to add volume to your soil, add compost or buy garden soil.

Water

I love a sunburnt country
A land of sweeping plains
Of ragged mountain ranges
Of drought and flooding rains

Those words written over 100 years ago by Dorothea Mackellar still strike fear into the heart of every gardener. Australia is the driest inhabitable continent on Earth and so water is something we are very conscious of conserving. The more than decade-long drought that affected much of Australia and resulted in critical water shortages across many capital cities, certainly did a lot to raise our awareness of water conservation. Around the world water supplies are being affected by drought and pollution, leading people in developed countries to place an increasing value on water. As a result of this we are seeing quite a shift towards xeriscaping, or dry climate gardening. We are being told to plant drought tolerant plants, and drought tolerance is becoming more common on plant labels in nurseries. This has been so drummed into us, that I have clients asking me to design them drought tolerant gardens even as we are receiving heavy rain.

Unless you live in an arid zone, you are likely to have some experience not just of drought, but also times when water is plentiful, even floods. Even in arid zones, floods can occur, but infrequently enough that you can safely call your garden a dry place and xeriscaping is perhaps your only option for a garden. In the subtropics where I live we certainly do experience dry periods, even extreme drought and water shortages at times, but between those times we can experience extreme wet. If you are not prepared for that cycle, those drought tolerant plants we were told we should plant can turn up their toes very quickly. Buying new plants every time the weather changes is not sustainable nor very realistic. In the tropics, subtropics and temperate zones, the wet will follow the drought, regardless of how many years that might take. Even in drought, tropical and subtropical zones will have high humidity. Many drought tolerant plants do not like high humidity and will dislike it even more when it is combined with high rainfall.

A 'one size fits all approach' to drought proofing gardens is not realistic. We know that both a lack of water and an excess of water are going to impact our gardens at some time. With climate change those impacts are intensifying and occurring in areas hitherto reasonably unaffected by weather extremes. My personal approach is to teach people to create resilient gardens - gardens that are able to cope with some degree of too much or too little water. Interestingly, the strategy for each extreme has a lot in common, and most of it is happening in the soil.

Drought proofing

Drought proofing a garden requires forward planning. It means planning for a lack of available water, whether that water is rainwater or another source of water. I will look at water sources in a moment, but first, let's look at some of the set-ups in the garden which will mean less water is required to care for the garden.

Soil – always start with your soil. As we discussed earlier, healthy soil equals healthy plants. Healthy soil has some built-in buffering against both drought and water logging. Organic matter in soils holds moisture, so by increasing the organic matter in your soil, you create soil which has a greater ability to hold water. Think of it like burying a sponge in your soil (but don't actually bury sponges, unless they are organic). If prolonged dry spells are on the horizon for you, go back to the chapter on soil and soil additives to work out the best way of improving your soil. Your aim is to increase the ability of your soil to hold moisture, without also making it susceptible to water logging.

A critical aspect in drought proofing your soil is to shade it from direct sunlight. This is most often achieved through mulching but can also be achieved through having sufficient plant cover to shade the soil. A densely planted garden is even better. This provides layers of shade that get deeper closer to the soil, allowing air pockets between the leaves which act as insulation. In such a garden, the air at ground level will be cooler and moister than a garden, even a mulched garden, where direct sunlight reaches the ground. Mulch does more than just shade and cool the soil below. It creates a layer to trap moisture as it evaporates, so is beneficial even for shady gardens.

In hot climates shade is important in cooling not just the soil, but also the plants. Cooler air holds less water vapour so evaporation of water from the soil, and transpiration of water from the leaves of plants, is slightly reduced in the cool of the shade. Some protection from direct strong sunlight during hot dry summer days will reduce stress on your plants. The term 'full sun' equates to six hours of direct sunlight a day. In many hot climates you can give your plants 'full sun' by lunch time and enjoy some shade for the coolness it brings to both you and the garden in the afternoon.

For long-term planning it may be beneficial to include some shade trees in your garden design. If chosen and placed carefully they can offer light shade that will still allow you to grow many of your sun lovers quite successfully. By placing a small tree on the western side of a garden or lawn, the area will still receive morning sun and possibly even direct sun into the early afternoon but will have some shelter from the harsh western sun in the afternoon. Be aware though of the natural shape and size of the tree and choose one that will only have the height and spread of the area you wish to shade. It is true that trees can create dry areas as they take more of the available moisture from the soil, but with good choices the benefits of the trees outweigh any disadvantages in a garden by a huge margin. The shade they provide can be enough to very significantly reduce the stress on your plants of a long hot and dry summer.

Hydrophobic soils

If your soils have become water repellent and water seems to simply run off the top as you are watering, you have hydrophobic soils and you need to act. Hydrophobic soils are soils which, instead of absorbing water as we expect them to do, will repel water, making watering the plants struggling to grow in these soils almost impossible. If you do not correct this problem, you can water and water and your plants will still die of thirst, as the water only runs off the surface.

Soils can become hydrophobic through prolonged drought, as the soil becomes so exceptionally dry and dead, it no longer has the life in it to interact with the water. It is far more likely to happen in soils with low levels of organic matter, and soils which are not mulched. It can also happen in soils mulched repeatedly with pine mulch or with eucalypt leaves. Both exude oils as they break down which can build up on the soil making it water repellent.

If you are not sure if this is happening in your garden, scratch the surface after watering – is it damp below? Or does the water sit in little pools surrounded by bone dry soil? There are a range of wetting agents and additives on the market which address this problem with varying degrees of success. Most of them can be avoided.

My preferred approach is to loosen the soil and dig a channel in it that can collect water before it runs off. Slowly water the soil with water from the bath, shower or washing machine – water which has a small amount of soap in it. The soap will help to break down the waxy coating that has developed on the soil. Next, spray liberally with a homemade compost tea, weed tea, worm juice or a seaweed extract which will further break down this coating and will add essential microbes back to the soil. Then add organic matter such as aged horse manure or compost, and mulch with straw or sugar cane mulch. Avoid using woodchip mulches at this early stage as they can add more oils, which only compounds the problem. The straw-based mulches will break down quickly, adding more valuable organic matter to the soil. Keep watering with mildly soapy water regularly until you are sure the problem is completely resolved. You will know you have got there when you see the water soaking in rather than running off, then dig a hole and see that the soil underneath is wet thoroughly. The same method, scaled to suit, will also work on large pots in which the potting mix has become depleted or hydrophobic.

The channel you dug will play a valuable role in slowing water runoff. Slowing the movement of water and giving it a chance to sit in one place, slowly sinking in, is vital to restoring hydrophobic soils, and is also beneficial in drought proofing. In areas of Africa where desertification has turned farmland into dustbowls, locals are having success simply by digging shallow holes through the barren earth. When rain does come, it is trapped in the shallow depressions rather than flowing away, and life can be restored. As gardeners we like to create nice level spaces in which to garden. Perhaps it is time to appreciate the role of a few dips and divots in trapping water in our gardens.

Water sources

Ok, you have your soil prepared to require less water, but we still need to think about where the water is coming from, if not falling from the sky. Most of us probably live in urban areas where we have access to municipal reticulated water. A lovely limitless supply of clean water at the turn of a tap, so long as we are prepared to pay for it, and that isn't usually much. We have learnt recently that this supply is NOT limitless, and we need to be careful with it if we want to keep it for things like drinking. As I write this in 2021 after a very wet couple of months, which included localised flooding in some areas of the catchment, Brisbane's water storage is at 67%. This is a combined result of twelve dams, about half of which are full. The largest and main dam, Wivenhoe, is still at only 40%. This highlights how variable our rainfall can be and why we should be prepared for almost any extreme with regard to water, even the possibility of future water restrictions.

As recently as 2010 Australia faced devastating drought. All major cities had water shortages and had to implement severe water restrictions. At the time Brisbane's water supplies got as low as 18% and authorities here implemented the toughest water restrictions that had ever been implemented by an urban water authority anywhere in the world. Australia became a world leader in water management and water treatment technology. Nothing drives action more than the threat of dry taps, so we all took on those restrictions with gusto, to the point that it drove permanent change to the way we value and use water.

For me it meant clean curtains. Washing curtains is usually so far down my list of things to do that it simply doesn't exist, but when I am not allowed to water my garden with town water, my water tanks are dry and I can still use my washing machine, use it I will! I washed everything I could to get water to recycle onto the garden.

I have a greywater system that waters my front cottage garden, so during the worst of the drought I would also encourage friends to shower at my house so that they helped water my patch. It did seem a little creepy I suppose asking friends to come over and shower, but gardeners will do what must be done to keep our plants alive!

In an urban situation the two most readily available alternative water sources are rainwater tanks and greywater. In a more rural setting, you may also have access to bore water, farm dams or have allowances to extract water from the local river.

Rainwater tanks

There was a time when councils around Australia were asking people to remove rainwater tanks as they were seen as ugly and possible mosquito breeding sites. How things have changed! New designs mean tanks can be less visible, located in tighter spaces, and even hidden under decks. During the drought, it became standard to include rainwater tanks on all new homes. With the drought over, this seems to have slipped out of mind. With the smaller size of gardens and the larger size of homes, there is less available space for tanks, making them less desirable. This need not be the case when you consider the options for tanks these days. They can easily be built under the ground in new developments so can be hidden even under the back patio. Garden spaces may be smaller but our need for water is not so significantly reduced. Washing cars, filling pools and hosing down of outdoor surfaces can all be huge water uses that can better be done with tank water.

The amount of water you can store will depend on how many tanks you can fit in. As a gardener I would suggest that the more the better. I have 7,000L capacity which usually sees me through a moderately dry winter and spring. They are often empty by the time the summer rain starts. A larger garden or one with a lot of tropical plants will need more than this.

Once the tanks are installed, this is free water, so make the most of it. There is no point having full tanks and gardens dying of thirst. Of course, the down side to rain water storage is that when there is no rain, there is nothing to store.

Greywater

Greywater is the water that comes from your bath, shower, hand basin and washing machine. In most cases it is perfectly fine to use on the garden, and it makes sense to redirect it out of council sewers and onto your garden.

Greywater does contain soap residues which can make the soil alkaline. Many detergents are high in phosphorous which will be damaging to some Australian natives. It will also contain pathogens which wash off your body. What does all this mean for the garden?

Greywater is best not used on acid loving plants. Apparently. I have been using greywater on my garden for 20 years and so far, nothing has objected at all, not even the acid lovers. Use eco-friendly soaps and detergents where you can, in particular ones which are low in phosphorous, and it will be fine for natives and pretty much anything.

Never spray greywater, always let it run directly onto the soil where any pathogens will quickly break down. Never use it on edibles – in particular, not on the fruits or leaves. And never store greywater, as to do so will cause pathogens to multiply and it will stink! Many councils regulate that greywater cannot be stored for more than 24 hours. I prefer to use a system that does not store it at all but allows it to run directly onto the garden. A healthy soil which contains organic matter and soil microbes will easily deal with pathogen issues. If you are washing a lot of dirty nappies however, it will be better to let this water go to sewer. Very few of us are washing anything with that extreme level of pathogens anymore, so it is unlikely that you will have any problems. Faecal bacteria break down within 24 hours in soil. They can survive for a much longer time in water. Make sure your greywater flows onto the soil and the soil is not becoming waterlogged, and you will not have a problem.

A note on temperature; hot water running directly onto the garden is usually not a great idea, unless perhaps you are growing warm climate plants in a cold climate. I tend to find that by the time my shower water flows down the pipes into the receiving vessel and then from there through the hose and onto the garden, the water is lukewarm at best and no problem for the plants or soil. The hose on my washing machine is short by comparison, so when I wash on a hot cycle it comes out very hot. My solution is to let it cool in a bucket before tipping it onto the garden. Obviously washing in cold water eliminates this problem (and saves electricity).

I also recommend that you take note of the bathroom cleaners you are using, as these too will go down the drain and onto your gardens. If you do like to use strong bleaches and harsh chemicals, use a lot of water to wash them down so they are very diluted before they reach the garden where they will harm soil microbes. Alternatively, divert your greywater to sewer during cleaning. All urban greywater systems should have a diverter switch which allows you to easily divert the greywater to sewer when it is not wanted or needed.

Using greywater on the garden requires access to your pipes which is not always easy. You could bucket water out of the bath, but I doubt many of us will keep up something that labour intensive for long. If you are building a new home, it could be worth investigating having greywater systems built in. There are some very sophisticated (and expensive) systems around that treat the greywater to some extent for you, but I am not a fan of these. They do work and work well, but why set up an input expensive treatment system for such a small volume of water when the municipal system is treating the water anyway? That being said, if you are a large household, it may not be such a small volume of water.

A very simple system is to put a two-way diverter onto your pipes, which allows you to direct the greywater into the sewer if there is wet weather and it is not needed on the garden. This makes it legal in most areas and is also very practical (BUT – do always check what your own council regulations are around greywater). Adding greywater to very wet soil contributes to water logging, which we don't want, and allows the pathogens in it to live in the water where they would otherwise die in the soil. Waterlogged soil creates a situation with free water for the pathogens to live and breed in which is not safe or healthy.

From your pipes the water needs to go into a holding vessel. This is because your pipes will be larger than the hose the water then flows into and by restricting the flow the water will back up. A holding vessel prevents it backing up into your bathroom. The holding vessel can be anything from an old wheelie bin, an old hot water container, or a plastic bin. The pipe goes in the top and the hose comes out the bottom. Very simple! As water does not sit in here, there is no potential for mosquitoes to breed in it, or for bad smells to take hold. A lid is always a good idea to prevent anything falling in, and to keep it nicely concealed.

An overflow is also a good idea if your vessel is not very large. Sizing your vessel to ensure it can comfortably hold one shower or bath is a good idea. If your household likes to shower in succession you may need to wait 10 minutes between showers for the water to run out or go for a larger vessel. In our house, we tend to have short showers and a plastic bin is enough. It does overflow sometimes and the Louisiana irises growing around it are very happy with that.

I move the hose around the garden most days to spread the water around, but you may prefer to direct it to run through a leaky hose or irrigation pipe to give broad spread of the water without having to move the hose regularly. Do keep in mind that the water is not pressurised and it is only gravity fed, so you will need to ensure the hose is downhill in the garden from the receiving vessel. Another bonus to a greywater system is that you can indulge in a wee in the shower to give your plants a very dilute dose of extra nitrogen.

Bore water

In certain areas you are allowed to sink a bore to access ground water supplies. You will need council approval to do this and that approval will regulate the size of the bore and how much water you can access. Groundwater quality varies enormously, so before you can use it for anything at all, even the garden, you will need to know what sort of water you are getting. The best place to start is find out if your neighbours are using bore water and ask them, as it will be the same underground supply you are tapping into. Some of the common problems with bore water are:

Temperature – some artesian supplies are actually hot – as in, hot springs! I had a memorable experience visiting a toilet in Winton in Western Queensland. I wasn't expecting a steam wash for my bum, but the toilets in the visitor centre were flushed using bore water that came up hot!

Salts – salinity of underground water is a naturally occurring problem that is becoming exacerbated by rising water tables in areas that are heavily irrigated. Highly saline bore water is not only undrinkable, it is highly toxic to your soils and plants, so be very careful with it.

Minerals – if your bore water is a funny colour or has a funny smell or taste, the answer is probably minerals. Sulphur (which smells like farts) and iron (which can make the water look rusty) are two of the commonest minerals found in high concentrations in bore water, but there can be others. Apart from looking and smelling funny, these are not usually a big problem for gardens. It is a good idea though to ensure there is plenty of organic matter in your soil to help buffer any changes this may make to your pH. You should also make sure you know which minerals are in oversupply so you can address any shortages in other essential minerals in your garden. If you find you have smelly bore water that is high in sulphur, grow garlic! The onion family need sulphur in the soil to develop their full flavour potential.

Contamination – contamination of groundwater is a serious and very real problem. In semi-rural areas it may be from unregistered or poorly performing septic systems and the contamination will be sewerage. In other areas it could be from mining or coal seam gas extraction, in which case the contamination could be acid, heavy metals, hydrocarbons or if in the vicinity of a uranium mine, can also be radioactivity.

To be on the safe side it is important to always have the bore water tested before you use it. If it is not safe to drink, think twice about using it in the garden, especially on edibles.

If the bore water turns out to be safe and clean, lucky you, you have a great water supply. But there is still a need to use it wisely. Underground water supplies are not inexhaustible, and they are important components of the natural landscape as well – many trees and even local creeks rely on underground water supplies. As the level in those supplies drop, tree roots can no longer reach it, and it can no longer feed into local creeks. When too much water is removed from any natural system, above or below ground, there will be far reaching ramifications, not just to the local vegetation and waterways, but also to soil salinity, local land quality and viability and the entire local environment.

Irrigation from local river systems

Taking water from local river systems requires a permit and is regulated. This is an area of much controversy here in Australia, given that our largest river system travels through four states and quotas are assigned by state governments. Water is taken by successive users without sufficient regard for the environmental impact or what may be left for downstream users. This is managed by the Murray-Darling Basin Authority. Within the catchment of this huge and complex system is Australia's largest irrigation area, and also Australia's largest privately (and foreign) owned dam (Cubbie Station, Queensland). No prizes then for guessing that this issue can be a political hot potato, and that the environment is likely to suffer because of it. Within this river system are many RAMSAR listed wetlands (listed as internationally critical for migratory birds) and many endangered ecosystems. Water misuse in this catchment has led to salinity problems and significant environmental devastation. These days salinity is monitored and managed, environmental flows are assigned and water allocations are a tradeable commodity. From crisis came management of this critical water supply, although the success of this management is questionable given that we see ongoing mass fish kills, and townsfolk fighting for enough water to even exist.

We have taken so much already from a river system, that we now need to go to the extreme lengths to determine how much water needs to be kept in the river for it to still be considered a river at all! I personally believe that here in Australia we have taken too much of our riverine and underground water supplies for granted. If you are lucky enough to have access to such water supplies, I urge you to use it responsibly. Many of you already do, as you rely on this water for survival, but for those who still wantonly splash it around instead of paying for town water – not cool.

Water wise gardening

Water wise and drought tolerant are NOT the same thing. Water wise means using water wisely – regardless of how much or how little we have of it. This means planning for water shortages AS WELL AS water excesses, if that is likely to be something our garden is subject to. Water wise is something we all need to be and is something we can even teach many of our plants to be. Sounds crazy, but as the chief water providers (in dry times anyway!) in our gardens, the way in which we water influences not only how much water we use but also how effective we are with that water, and how needy our plants are for water.

When we water the garden deeply, we wet the soil profile to at least 10cm deep. In terms of the root systems of most trees, shrubs, and even perennials and annuals, this is not actually very deep. When we water in a hurry, and this includes most hand watering, we often only water to two or three centimetres deep, or less. Plant roots will seek out water, and so will be strongest in the soil that holds the right amount of water. A good deep soaking (watering at least 10cm deep) once a week will allow the surface to dry out a little, and for the plant roots to travel deeper into the soil seeking water. These deeper roots are then in a cooler and more consistently moist zone in the soil, making them less heat stressed, and less water stressed.

Realistically we want our soil to hold moisture much deeper than 10cm. A good drenching by us or nature will wet the soil profile thoroughly. Once the soil is deeply watered, by watering the top 10cm regularly, we are keeping the entire profile moist, as capillary action will draw the water we have added deeper into the soil. This cannot happen if the layer of soil below where our water has penetrated to is dry.

Consistent shallow watering will result in roots staying close to the surface as this is where the water is. It is also the hottest part of the soil, and the area that can dry out and scorch the roots very quickly. I have known gardeners to kill plants through regular very shallow watering. One client of mine has killed an entire garden worth of plants many times over this way. I hope one of these days I will succeed in convincing him that he is wasting water and wasting plants. A deep soak once a week is more water efficient than small amounts often, and your plants will be stronger for it. Unless! If you are growing orchids, ferns, bromeliads and many of the Gesneriads (African violets and related plants usually favoured by advanced and specialist gardeners), the opposite is true – water small amounts often! As soon as someone asks me why the bromeliads or ferns are dominating in their garden and they are struggling to grow anything else, I know we have a problem with frequent shallow watering.

If you are unsure of how effective your watering has been, dig a hole. This is the most surefire way to understand your garden, your soil, and your watering practices. I am not joking – get out there and dig a hole and find out what is happening in your soil! Amazing how much you can learn through just digging a hole and looking at the soil. How deep has the water penetrated into the soil profile? If 'not very' is your answer, get out and water again. Similarly, if we have had rain, or even a storm, it may be wonderful but not deep enough. This is a good time to accept the help from Mother Nature and add some extra watering to ensure the soil profile is watered to at least 10cm, or even better, 20cm or more.

The deeper the water is in the soil profile, the longer it will take for the lower depths to dry out. This encourages roots deeper into the soil, making plants stronger and more drought hardy. Keeping in mind that for an area to be officially drought declared, the requirement is for there to be no plant available moisture in the top 60cm of soil. Not many of our gardens will really withstand that, and it goes to show the importance of deep watering and deep-rooted plants.

Another means of reducing water wastage, and at the same time, teaching plants to be water wise, is to only water when the plants need to be watered. Most plants and gardeners alike will wilt in the sun on a hot day. Once the sun is gone and the cool shade returns, do the plants pick up again? If they stay wilted, they could do with a drink. If they pick up again, hold off the water, they don't really need it yet. Sure, the extra water might be great and make them bigger quicker, but it won't make them stronger, and if water is in short supply, it is wasteful. Conversely, if we do not water plants when they are telling us they need it (e.g., remaining wilted into the cool evening or early morning), they will become stressed, weak and possibly die.

Another aspect of being water wise is the timing of watering. As a general rule, water in the cool of the evening. At this time the water stays in the soil longer and therefore seeps in deeper as less is being lost to evaporation. During the day when the sun is hot a significant amount of water evaporates before it gets the chance to penetrate into the soil and do any good. Over the cool of the night and into the early morning, plants are able to utilise that water before they too start to lose it through transpiration.

When temperatures drop below around 15°C at night, switch watering to the mornings. Plant activity slows rapidly at this temperature and the plants are unable to take up and utilise the water during the cooler nights, making them more susceptible to fungal attack.

We are often told that plants, such as roses and cucurbits, which tend to be susceptible to black spot and powdery mildew, should be always watered in the mornings. The important thing here is to prevent water sitting on the leaf over the cool of the night. This can be minimised by watering by bucket to miss the leaves, or by drip irrigation. Of course, rain is not going to miss the leaves, so watering is only part of the story with leaf fungal diseases. A soil with good organic matter and microbes will help to provide the beneficial bacteria and fungi which destroy the bad ones.

Irrigation

I like irrigation, in most cases. If used properly it can ensure we give things the deep drink they need, as discussed above. If used badly it can be very wasteful. Mind you, sprinklers are the same – use them well and they do a good job. During water restrictions, when we were not allowed to use sprinklers, but we were allowed to hand water, many plants were lost to regular shallow watering which left the roots near the surface of the soil. Admittedly this regulation was mainly targeting lawns, which will brown off badly in drought but come back bright and green very quickly when it does rain. Lawns are particularly susceptible to die from shallow watering, so better to let them brown off if you are not able to give them a proper watering.

Back to irrigation. For purposes of sustainability, we are told to save water by using drip irrigation installed under the mulch. This does water the soil directly and save on water lost to spray drift so is a good way to use water efficiently. It is not always the best answer. There are times when you do want water coming from above.

Gardens full of epiphytes such as orchids and tillandsias will need water delivered to them in the trees. These gardens are usually in humid climates and have shade, so the amount of water lost to evaporation will be less. Still, the watering system should be used carefully to avoid the heat of the day and wind. If you want to use this system, choose spray heads which have a reasonably large drop size. The larger drops are harder to blow away and are far more effective at penetrating the mulch yet are still able to contribute to humidity in the garden. Spray heads with a fine spray are designed for use in a closed environment like a green house. In the garden the fine mist is almost completely lost before it reaches the ground, and if it does, the drops are too small to be able to penetrate the mulch. Misters are usually a smaller and therefore cheaper spray head, so get mistakenly used in situations they are not designed for.

Drip or trickle irrigation, which is usually placed underground or at least under the mulch, will not carry nutrients from the surface down to the soil. If you are fertilising or even adding compost, you will need to rely on rain or hand watering to wash the goodness down into the soil. Nor does drip irrigation wash dirt, dust and pollutants off the leaves of plants. This is not a 'biggy' if you get periodic rain. If you don't, or your plants are undercover, the plants will thank you for the occasional wash down. A build-up of dust and dirt on the leaf will reduce light hitting the leaf surface, reduce their ability to photosynthesise and will block stomata, reducing gas exchange. In general, the larger and greener the leaf, the more this will be a problem for them.

Keep any irrigation system simple. You will need to be able to turn it off to prevent over watering in times of rain or if watering in fertiliser. In reality the right irrigation, and therefore the most sustainable irrigation, will be the one most suited to your garden and the plants you are growing. Fit for purpose must be a consideration in any irrigation system if you want it to do the job effectively and with the greatest water efficiency.

Too much water

Some level of drought protection is important, as is some level of improved drainage for wet periods. New housing developments are often occurring on marginal lands – lands that were once wetlands or floodplains. These areas may well be drained and built up and even have state of the art stormwater drains with backflow devices fitted, but there will come a time when there will be heavy rain which will cause waterlogging even if you escape flooding.

Significant rainfall events which lead to localised flooding are common in the tropics and subtropics but are becoming more common all over the world. Weather extremes are occurring more often instead of nice predictable rainfall patterns. Too dry or too wet are becoming normal. Sustainable gardening talks a lot about drought proofing gardens, but we equally need to allow for managing too much water in our gardens.

If you have sandy soils, it will be less of an issue for you than those of us with clay soils, but if you also have a reasonably high water table, which occurs in many developments on reclaimed wetlands, your sandy soil can waterlog as well.

Waterlogging occurs when the air spaces in your soil are filled with water. A healthy soil has lots of air spaces and the decomposition process happening in these soils remains aerobic. One of the key jobs of worms (who breathe air) is to leave tiny tunnels in the soil which keep the soil well aerated. Compacted soil is soil that has been squashed over time, by foot traffic or other activity, by heavy rain pounding it, or just by the heavy nature of the soil. Compacted soils are hard to dig. The reason for this is that the air spaces have been forced out of the soil. Plants need to have their roots in contact with soil rather than air pockets (hence the requirement to water in new plantings), but that soil does still need to have tiny air spaces in it. The process of plant roots taking up water and nutrients in the soil needs to happen in an aerobic environment (e.g., one which contains oxygen). Plants which grow in bogs and wetlands have special adaptions to allow them to grow in soils which have low oxygen levels – which is one reason why they are a specialised group of plants that rarely do well in a normal garden environment. Conversely, plants that are not adapted to grow in boggy soils do not like it and will 'suffocate'.

Waterlogged soils tend to drown the worms and microbes in the soil, leaving the soil devoid of life. You will notice that waterlogged soils develop a sour smell. This smell always indicates a lack of oxygen, whether that be in the soil or the compost. It means that the active bacteria are anaerobic bacteria (living without oxygen) which, whilst still continuing the breakdown processes, are not supplying nutrients in a form most suitable for plants to use.

So, what do you do if your garden has a tendency to waterlog? To begin with, you need to decide how regular an event this may be. Periodic flooding will be managed very differently to regular or even semi-regular waterlogging.

The very first step is soil improvement. This will assist with drought proofing at the same time. Check back in on the previous chapter about soil if you are still unsure about how to do this. In relation to waterlogging, added organic matter in your soil will open up the soil, making it more porous and creating more air spaces, which means it can absorb a lot more water before it reaches a waterlogged state. If your problem is not major, this may be enough.

Often new homes are built in such a way that the underlying clay is left at the surface, causing problems. By adding organic matter (and possibly some gypsum or dolomite, or even better, rock minerals containing silica) you can break this clay up and get some decent soil again. In this case the waterlogging problem is a simple matter of dealing with the clay which is no longer as deep as it should be. By making it more friable and creating decent garden soil out of it, the waterlogging problem can be improved, possibly even solved.

Creating mounded or raised gardens with good quality soil will also help to ensure plants have their roots above the waterlogged zone in the soil. In severe cases or if you are growing plants that particularly hate wet feet, garden beds that are built over mounds of gravel or rubble will be much more free draining – not so good if water is in short supply but great when there is too much of it. Building a raised garden bed over a large mound of sticks, logs and organic matter is a good way of creating both a temporary and a long-term solution. In the short term you will have decent soil to plant into which will be free draining. Over time the buried organic matter breaks down and helps improve the heavy soil below the bed. As this happens the mound will sink, so the raised bed will not remain raised.

Water logging may also be caused by poor drainage due to the lay of the land. If the water cannot escape, the result will be flooding to some degree. As I write this, we are in the midst of a three-day heavy rainfall event expected to drop approximately 250mm over South East Queensland, coming after a very dry summer. My garden is already a series of very large puddles which are quickly joining into a backyard lake. The only real solution is to find ways for the water to escape. This could be as simple as digging a channel, or as complex as major earth works.

Let's look at the simpler solution before we go and book an excavator. A channel dug in the dirt to allow the water to escape to lower ground will work in some cases, and sometimes is all that is needed. For the sake of neatness, or if large volumes of water are to be channeled, it is preferable to make the channel more substantial. This usually means a trench approximately 30cm deep filled with drainage gravel. Drainage gravel has a size of around 5cm. Any smaller and the gaps between the gravel will be smaller and less water will be channeled through it. Larger gravel will have larger gaps which tend to clog with dirt more easily, again meaning less water can pass through.

A trench filled with drainage gravel should usually have some fall (slope) away from where the water begins towards where you want the water to go. If the distance is very long (e.g., greater than 30m) or includes lots of bends, you may find that laying ag pipe in the bottom of the trench then backfilling with gravel will be more effective at directing large volumes of water away.

A drainage trench that has nowhere to go can be directed to a drainage pit. A drainage pit is a large hole, usually a metre square or larger, and filled with drainage gravel the same as the trench. This can be a very effective way of handling medium amounts of water if there is no other drainage point, but once the drainage pit is full, it will overflow if water enters it faster than it can seep into the subsoil.

This brings us to a point where you can use waterlogging measures to also help with drought proofing. A drainage trench or pit will not just direct water away from where you don't want it, it can be used to direct water to where you do want it. In the simplest form that is lower in the soil, leaving the surface less impacted.

I have numerous drainage trenches in my garden, as I have a low-lying block with heavy clay soil. I have one small storm water drain, and two places where excess water can leave the property into a neighbouring easement (until that also floods, which is often). I have made sure all of my trenches are deeper than their exit point. This means that water channelled into the trench will remain in the bottom and provide a deep watering for nearby plants and trees before the excess is channelled away. This method can save the surface layers of the nearby garden bed becoming waterlogged, but in times of drought when less water is available, any rain we do get is prevented from running away too quickly. Whilst these techniques will save most of my garden from complete waterlogging, it really only protects the top 20cm, below this the soil can still waterlog which will certainly affect deep-rooted plants.

Planting trees is very important in dealing with water deeper into the soil profile. The deep roots of trees will help draw water from the lower profile of the soil, helping to keep water moving deeper into the soil. Areas which are prone to seasonal inundation or waterlogging tend to have a lot of deep-rooted trees. When these trees are removed for development, water is no longer being cycled through as deeply in the soil and the water table often rises, making waterlogging a bigger problem.

A gravel trench can be hidden underneath the lawn or garden bed, where it can work to direct away excess water while also trapping small amounts of water and increasing deep watering. In order to do this, the outflow point of the gravel filled trench should be higher than the bottom of the trench, so that excess water will flow away after the trench fills.

If none of these solutions have resolved the problem sufficiently, you may need to get more drastic and look at earthworks to adjust levels and create more significant drainage. Depending on the extent of the earthworks, you may need council approval. In any case if you are directing water off your property it needs to be done in accordance with regulations, and with consideration for who is likely to be receiving that excess water. It is most unneighbourly to create water problems for others.

If all else fails, perhaps it is time to simply embrace the wet conditions and grow plants that will love it. There are some very attractive plants that grow in natural seasonally inundated soils that handle both waterlogging and dry conditions. Think day lilies, canna and Louisiana iris. Just check first that they belong in your climate zone for best results.

Emma's Garden - Taming a challenging landscape

The first sign that there was a garden of interest here was the huge tree in the front garden and the verge planted out with natives. It stood out in a street where gardens are fairly minimal and lawns are favoured. Emma describes herself and her husband as aspirational gardeners. She doesn't think she knows what they are doing. Her husband Chris is more confident and has a very clear idea of what he is doing in the garden!

Ten years ago they moved into this house on a sloping block. It didn't take long till they found that stormwater from their own and other gardens ended up under their house. Any muddy storm water that didn't stop there ended up over the road under neighbours' houses and in neighbours' swimming pools. For the sake of their own liveability something needed to be done.

Emma was adamant that they needed to keep as much of that water onsite as possible. While she envisaged natural pools and watercourses crossed by wooden decking, Chris took a more pragmatic approach to create a very beautiful but also highly practical garden. Landscaping was needed.

To create functional flat areas, a terraced area was created. This allowed for a small flat lawn which is a lovely feature and great play space, but also works to allow much of the stormwater to soak in before running off. Below the lawn is a short but steep bank. It is held together with coir logs and densely planted with native trees and grasses.

This couple very much wanted a native garden as they wished to conserve local native habitat and support wildlife. Happily, in setting up the garden they went to a local native nursery which sells only native plants indigenous to the greater area. They did not just get plants here, they got good advice. Following the advice, they used lots of lomandras, native sedges and grasses for erosion control and to slow down water runoff on the steep slope below the lawn.

At the base of the slope is a beautiful rocky creek bed with ag pipe below it, which wraps around the house, to the side garden where it skirts the edible garden beds and finally to stormwater at the front of the house. Ag pipe has slots in the pipe which allow the water to escape from the pipe into the gravel bed. This means that when the water flow is light, it will all escape the pipe into the garden where it is needed. When there is a large downpour and water flows are high, excess water is directed to the stormwater without scouring the surface of the garden bed.

They no longer have issues with storm water flowing under their home, nor are they contributing to downhill problems for other neighbours. In fact, they have been so successful in trapping water on site that they now have damp patches which they can easily address through improvement of the heavy clay soils below, and through careful placement of plants which like lots of water into those spots.

Behind the lawn area is a fantastically huge chook pen on one side, and a concrete pad surrounded by a short rock retaining wall. Emma very much wanted a softer surface here such as timber decking or decomposed granite. Chris won this debate on the grounds that the timber would need replacing in time and the decomposed granite would get muddy in our subtropical summers and rot the table they wanted to put there. Chris has taken the approach to landscaping that it should be done once and built to last. This approach has meant that materials used are fit for purpose with a long lifespan. This concrete pad is anything but a boring area. It has been pebblecreted to give a more natural surface which is less likely to get stained by leaf fall from the surrounding trees. It is very shaded so does not contribute to radiant heat issues. Water running off this hard surface flows directly to the garden bed in front of it where it waters trees. This area is now a well used entertaining space with table and chairs and a fire pit.

The garden features a huge number of native trees, not all of them planted by Emma and Chris. The huge lilly pilly in the front garden was here when they came, as was an old tea tree at the back. This old tea tree was smothered under a neighbouring poinciana which is now gone. It can now grow upright, and is doing so with gusto, but has retained a gloriously tortured shape which is now a real feature in this part of the garden. Many gardeners would have cut the tree back hard to improve the shape, but this tree is enjoyed for the beauty and the story of the untamed shape.

Nearby are a couple of volunteer trees, a golden penda and a Norfolk pine. Norfolk pines can be very difficult to garden under due to allelopathy but this one is small enough that other trees and plants are able to get established nearby before that becomes an issue. As there are a few remaining large old Norfolk pines in the district which are part of the history of the area, keeping and enjoying this tree feels locally significant.

This tree loving couple have added numerous more native trees including small trees such as leptospermums, banksias, casuarinas, and even a Queensland bottle tree, combined with larger rainforest trees like the fire wheel and a stunning blue quandong. There are also a couple of peach trees that were here when they bought the home. One is about to come out to make way for the new mango tree, the other is about to be pruned very hard to allow them to manage the fruit fly by netting the tree.

In addition to natives, edibles are high on the list of desired plants in this garden. There are a number of raised timber vege beds which are currently being used as respite care for plants waiting for their new home as the front garden gets reworked. The inground vege beds are still providing food however with self-sown tomatoes and herbs. Weeds and scraps are all fed to the chooks who return the favour with eggs. This is one the best built chook pens I have ever seen. While it is a simple design it is sturdy and strong and like everything Chris does, it is built to last. The chooks are safe in there, and it must be cosy for there is often a possum sleeping in there as well. A resident python keeps the mice under control.

The garden is also home to a couple of native bee hives, and an energetic dog, Lucy.
This garden is very much a work in progress. With the landscaping finished at the back, they are now looking at refashioning the front garden to manage water there and to increase biodiversity. There are two large leopard trees on the footpath, which are not native and have high weed potential. Removing large trees of any sort is not good for wildlife so Emma and

Chris have taken a long-term approach to replacing these trees with native trees, by placing strangler fig seedlings into the forks in the leopard trees. The figs will gradually take over the leopard tree, so that the unwanted tree is eventually replaced by a native tree with much greater benefit to wildlife.

Long may this couple garden! If only more people were this responsible in managing stormwater flows on their property, and so interested in supporting local biodiversity.

Landscaping Vs Gardening

Landscaping and gardening are two different things. Landscaping usually involves putting the hardscaping first and plants come later. Gardening pretty much revolves completely around the plants, and hardscaping is done to enhance the role of the plants.

This book is written for the home and professional gardener. It is still entirely relevant to the professional landscaper however I do encourage landscapers to seek out additional publications which will go into more detail relevant to sustainable landscaping on a professional scale.

Hardscaping and hard surfaces

Hardscaping refers to the creation of hard surfaces in a landscape. This can take the form of retaining walls, driveways, patios, pergolas, paths, decks and any other hard surface used in the garden setting. These features can make a garden a very comfortable and attractive space to be in, can greatly assist with access and can be vital if accommodating a large number of people in the garden. However, there are downsides and one of these is the expense – the materials and effort required can be considerable with commensurate costs. With material use on any scale comes the issue of how sustainable those materials are, which is covered in depth below.

Possibly the two most significant sustainability aspects of hardscaping relate to water penetration and heat storage; that is, the way the materials allow water to either penetrate or run off, and how those materials affect the temperature of the garden through storing and radiating heat.

Water penetration

Water runs off hard surfaces, such as concrete or paving. It soaks into ground that is covered in plants, lawn or ground covers. Water that is not absorbed into the ground needs to go somewhere, usually the stormwater system. With increased loading on the stormwater system comes increased local flooding, increased pollution of local waterways and increased localised drought conditions. This can be a citywide problem as well as a problem in our own garden.

The additional runoff created by hard surfaces needs to be considered. Where will it go? Does the space slope towards your home, towards the garden beds or off site? Storm water runoff will carry with it anything that it can pick up – pollutants and weed seeds included. With the increase in runoff comes the increase in pollutants such as fertilisers, weed sprays, leaked petrol, flecks of rubber from our tyres. Very little of what builds up to toxic levels in local waterways is actually visible to us. We can't see what is being washed away with the stormwater, so we need to be aware of what potential there may be from our site.

I also mentioned the exacerbation of local drought conditions. When you have only a small area for water to penetrate, the ground next to this, which is covered by concrete, remains dry. Water seeps through the soil, so effectively you have less water leaching into the same area, creating drier garden conditions. This can be overcome by using porous paving, or gravel instead of a solid surface. If you do need a solid surface, perhaps you can slope it slightly towards the garden bed and put a small gravel trench between the solid surface and the garden. The trench will catch the runoff and channel it deeper into the soil so that it does not cause surface flooding of the garden bed.

Increasingly town planners are using porous paving around street trees. It has become well documented that these poor trees planted into footpaths with tiny little surrounds are starved of not only water but air as well. Soil needs to breathe, and trees need space in the soil to spread their roots. A little hole punched through a concrete footpath may fit the tree's trunk but does not allow for what needs to happen below ground in order for the tree to thrive.

Radiant heat

All hard surfaces absorb, store and release heat. Even bare dirt. Living green covers are significantly cooler surfaces. A lawn will always be a cooler surface than a paved area. While the surface temperature could be massively different, the ambient air temperature can be as much as 10°C cooler above a lawn as opposed to a hard surface. As the amount of living green increases, the cooling impact increases. A shrubbery is cooler than a lawn and a shade tree even cooler again. The natural transpiration of plants is part of the cooling effect. Shade, and the reduction of heat absorbing surfaces is also key. The heat absorption of hard surfaces is causing the heat island effect we talked about earlier. Reduction of hard surfaces and a corresponding increase in green space, even by way of green walls is currently a key strategy many municipal authorities are implementing to combat the effects of climate change. We can apply the same idea in our own back yards. More green, less hard surface will be cooler for your personal living environment and can go so far as to provide significant passive cooling of your home.

On the subject of transpiration – the increased water vapour in the air due to transpiration from living plants is a major contributing factor to local rainfall. Rainforests can create their own climate including a cycle of constant regular rain based on the high levels of transpiration. Conversely, deserts are able to do the same through their extremely low levels of transpiration leading to a lack of rain. Sure, there are other major climate factors affecting rain but transpiration is still very important. In the concrete jungles we are creating to live in, we are creating ecological deserts without transpiration.

When rain is very localised, it will often follow corridors of vegetation before it will follow the concrete jungle. Plant a tree or two and be part of the green corridor which might just attract more rain. At the very least let a tree shade the concrete driveway and prevent it from heating the house. Even better – use porous paving or gravel for the drive and allow the water to penetrate and water that shade tree. It will grow better and give more cooling shade.

The upshot of all this is that any hard landscaping you use in your garden will have more impacts on your livability and the health of your garden than just the aesthetics of the space. Think about how you can create shade over your patio, or how to direct the rainwater runoff to your garden beds, and that same patio will be cooler, greener and a much happier space for you and your plants.

Landscaping materials

Recycled materials are definitely a sustainable option, reused or repurposed is even better, but not always easy to source. We all love the pictures we see of gardens landscaped using salvaged wharf timbers with rusty metal to decorate it, but few of us are lucky enough to have access to these materials. Decking made from recycled plastic looks good and is durable but is expensive. It also doesn't have the warmth of wood, and for that reason alone there has been some hesitancy in using it.

Ultimately the most sustainable thing to do is limit the use of materials full stop. That includes materials used for hard landscaping, and all the smaller materials used in the garden such as pots, water features, irrigation, lighting and ornaments. Sounds rather boring doesn't it? We do all love to see beautifully landscaped gardens, but we need to consider how much landscaping is actually necessary in creating our dream garden.

Thirty years ago, the average backyard looked rather different to today. It might have included a bit of a porch or patio, and then was mostly grass with some shrubs around the edge. These days, the garden is smaller, there is less lawn as kids play there less, and more landscaping to enhance the smaller space. This may include outdoor seating areas, paths, retaining walls and raised gardens, pools, decking, fire pits and BBQ spaces or even just the tendency to concrete an area because we have no other use for it and don't want to worry about mowing or weeds.

In addition to the problems of lots of hard surfaces discussed above, this is all resources and money. If we do wish to include areas of hard surfaces, we should try as much as possible to make them porous. Gravel is a good option here for paths or those awkward spots that we don't want to mow as it allows water to penetrate. It will store and reflect heat though. If you are using paving, try using unsealed porous pavers which will allow some water penetration, or lay them with gaps that can be planted with groundcovers or low growing plants. This breaks up the large surface, creating spaces where water can penetrate rather than running off. The extra green also helps reduce the reflected heat from the surface.

Hard landscaping materials also tend to be expensive and have high intrinsic energy through mining, processing, manufacturing and transporting, so for the sake of limiting the overall resource use of our gardens, are best kept to a minimum. If we have access to recycled or reused materials the intrinsic energy of those products is lower, and we have managed to keep something out of landfill which is always a plus when talking sustainability.

Old railway sleepers used to be a common way of using recycled materials. These days modern railway sleepers are often concrete because they last longer, so there is a much shorter supply of the true recycled product. Instead, most sleepers used in landscaping are sourced directly from plantation timber and are a new product. Look for 'railway sleepers' rather than simply 'sleepers' when looking for a secondhand product. Railway sleepers will also have the marks of age and wear which gives the additional character to recycled timbers which we love, and which is another good reason to choose a recycled product. In fact, we love it so much, these beautifully aged timbers can now fetch much higher prices than new timbers.

We are seeing a huge rise in garden furniture and garden art being produced from old metal and timbers. We are equally seeing a rise in brand new items being mass manufactured and then treated to look aged. Recycled pieces tend to be one-offs, due to the limited nature of the materials used, and are often more expensive due to their handcrafted nature. Investing in a beautiful piece of recycled garden furniture or art can be expensive but a worthwhile way to make a statement in your garden. Your piece will likely be very unique, will have less travel miles as it is probably locally made, and will also have supported a local artisan. You may also be lucky enough to find out a little about what the bits of metal or timber came from and there could well be a worthwhile story there too.

But – back to the hardscaping part. Using reclaimed materials can be challenging. One of the main reasons is getting enough of what you need in the same size, colour or pattern. Online trading sites have facilitated access to reclaimed materials. Much of this is driven by someone not wanting the effort and expense of disposing of their old pavers, and so offering them free or cheaply online.

Reclaimed = previously used and now to be used again in a different setting but for the same purpose as they were made for.

Repurposed = using an item that has previously been used for some purpose other than what it was made for.

Recycled = breaking down an item into components which can be remade into a completely new product.

Materials such as pavers, old bricks, gravel, sand, rocks, timbers and fencing are all commonly available for free or cheaply on these sites. We tend to be selective about choosing what suits us, and wait for the right-coloured pavers, in the right amount for example, and in a location that is easy to collect from. Because of this it is often easier to go to the landscaping yard and get exactly what you want as you want it. We humans do like to go for the easy option.

Using reclaimed landscaping materials can take longer and require more patience to acquire. We may need to be more flexible and creative with our design to allow for what we can actually get rather than what we may think we need. Reclaimed materials can be a bit trickier to work with. On the plus side, they are usually much cheaper, or even free and the end results can be far more creative and more satisfying. And it is better for the environment.

Landscapers don't get a lot of opportunity to use reclaimed materials as they are working to deadlines and negotiating the specific demands of clients. It can also take longer to work with materials that are not ordered to measure, so if you are working with a landscaper, expect them to charge more for the privilege. Paying more for labour might still work out for you if the materials are free.

For most of us who are taking the gardening first approach, using reclaimed materials is less difficult. We can do it bit by bit over time. We can change our minds and rearrange it all on a whim, or as the garden grows.

My own garden has been almost entirely landscaped with recycled materials – mostly without plans but being designed on the hop as things come to hand. This has led to an ever-changing and ever-improving garden which is interesting and satisfying. My garden includes rock edges from a couple of different gardens that were getting rid of rocks, white gravel paths from gardens that were being re-landscaped and getting rid of gravel previously laid, old brick paths sourced from rubble on demolition sites, a timber boardwalk sourced from an old hardwood fence that came down. Even other garden features such as ornamental pots and planters, a shade house, garden ornaments, my potting bench, garden furniture, gates and a slide mostly came from unwanted items to be taken from a client's garden or from kerbside rubbish collections. I have even sourced plants this way. My son's cubby was built entirely from recycled pallets and the chook pen from an old shed and recycled wire. As I said, all very satisfying, but it does take time, and this is a garden that is so far over 20 years in the making and remaking.

Where we don't have that sort of time, patience or even the expertise to work with materials which are not an exact fit, using reclaimed materials gets much harder. If we can use them for just one element of our landscape, we have successfully cut down on the amount of new resources needed for our garden project. Sometimes it is just a matter of making the small changes that work for us, rather than trying to go the whole mile.

There are some recycled materials that are available in bulk supply from landscapers. These include gravel made from recycled concrete, and timber decking made from recycled plastic. You will even find garden furniture and edging made from recycled plastic, and some of it is very good. These products are true recycled products whereas the materials we just discussed are actually reused or reclaimed materials – that is they are used a second time around in the same form.

When a product is recycled it is changed from one form to another, and this takes energy. Reused is obviously the best way to go as no further energy is required. Recycled products usually require significantly less energy and resources to create than new products and have the added bonus of keeping good materials out of landfill, so while there is still an energy input, they are far better environmentally than using new resources or first use products. Repurposed products are those for which the second use is not what it was initially created for, and while it may need some modification, it needs no re-manufacturing to create it for its second purpose.

So far we have looked at limiting hard surfaces, and using reused and recycled materials in order to increase the sustainability of your landscaping. Where materials do need to be brought in, understanding a bit about those materials is worthwhile. All materials used in landscaping have an eco footprint.

A smaller eco footprint may be our ultimate goal, but if the products we choose don't last and need to be replaced regularly, we aren't achieving this goal. Fit for purpose is a strong consideration, and possibly the most important aspect when deciding on materials. As with so many things, if our landscaping is built to last it may cost more upfront both financially and energetically, but the savings are made in the long run.

The power of consumer choice also plays a part in choosing landscaping materials. We can choose not to use imported materials when there are great local alternatives. We can choose not to use timber with toxic treatments, and we can even look for companies with eco-credentials when choosing suppliers, materials and landscapers. This process all starts with asking a few more questions when we start planning our project.

Timber

Timber is a natural material, and technically is renewable. It is however renewable only if it is sourced from plantation or regrowth forests, as opposed to virgin forests, and if new trees are being planted to provide future timber supplies.

Timber is attractive to termites and has a tendency to rot, both of which greatly reduce its usable lifespan in a garden. Different tree species have different properties, so it's critical to think about the type of timber that is fit for purpose when it comes to using it in your garden. Basically, timber is divided into softwood or hardwood, and is available as a treated or untreated product. The softer the wood, the shorter its lifespan. Treated wood will last longer than untreated timber, but a treated softwood like pine will still have a shorter lifespan than an untreated hardwood.

Timber is treated in a number of different ways, but most commonly it is using a copper and chrome solution. While this is nowhere near a toxic as the old version which included arsenic, it is still a consideration if you are using the timber to build a raised vegetable garden for example, as the chemical treatment will leach into the soil to some degree.

The bulk of the landscaping timbers available in Australia these days are sourced from plantation timbers. If in doubt, ask your supplier. They may not know where the timber comes from but should be able to tell you if it is FSC certified or not. FSC stands for Forestry Stewardship Council, an international body which certifies timber and timber-based products (including paper) as being sourced from sustainably managed forests.

Specialty timbers are not usually used in landscaping as they are too valuable, so are reserved for furniture or flooring. Garden furniture is an area that uses specialty timbers. It is also an area where much is imported from parts of the world with very damaging forestry practices.

Rainforest timbers from South East Asia make beautiful garden furniture and many are naturally rot resistant so perfect for outdoor use. Sadly, many are also illegally harvested from critically endangered habitat, which is not so appealing. Again, ask the supplier if your timber garden furniture is FSC certified. If they cannot tell you it is, don't buy it. There are always other options, so this is not an area where it is hard to make conscious consumer choices.

When it comes to the project you have in mind, these days we should also be asking if timber is indeed a suitable choice at all. Timber is beautiful and is almost always the more attractive look in a garden, but if it does not last, or is going to leach toxic chemicals, it is not such a good idea.

If you are building a retaining wall with sleepers, go for the concrete sleepers that are designed to look like timber. They are not as beautiful, but they will last forever. We usually want retaining walls to last a long time, so it makes sense to build them using materials that will not need replacing in 20 years' time.

I am currently repairing my 25-year-old timber fence. It will have another five years in it if I am lucky. It gets a lot of wet soil against the bottom of it during heavy rain, so its life was always going to be limited. When the time comes to replace it, I would certainly prefer the look of a timber fence, but I need to build a brick base for it to prevent it rotting at the bottom next time. If this makes the timber fence last much longer, it will be worthwhile. In other parts of the garden, a 20-year life span may be all that is needed.

Gravel

Gravel is another landscaping product that is versatile, popular and comes in a great variety to choose from. All gravels are quarried, which is not good for the environment. Many are locally quarried, usually as part of local development. The very localised nature of this industry is certainly a positive.

My own garden was once a quarry. This site was quarried to provide rock to build an overpass over a railway line less than 100 metres away. After quarrying, the site was filled with construction waste and covered with subsoil (mostly clay) excavated from another nearby development site. This has given me quite a journey from a gardener's nightmare to decent soil, but it is a local story.

Any gravel you purchase from a landscaping yard is likely to be locally quarried. If you are using drainage gravel and it is to be buried, choosing a cheaper product will probably also be a more sustainable choice. It is probably recycled concrete. This is a demolition waste product. Given the high level of energy, water and material inputs required to make concrete, it is great to see the shift towards recycling it as much as possible. Smashing, screening and tumbling it to make gravel is a great option and is very likely to be how your drainage gravel came to be.

When we start wanting specialised gravel finishes, the sustainability factor drops fast. Black or white gravel is unlikely to be locally quarried, and there is a high chance it has come from a very unsustainable mine in the third world. This also goes for all the fancy bagged gravel you buy for little garden features. Basically, the less natural your gravel looks, the less sustainable it is.

Pavers

We can buy good old standard locally made concrete pavers, or we can buy very fancy fashionable pavers imported from somewhere else - or something in between. The more fashionable our pavers are, the less timeless they are likely to be. Fashions come and go! Will you be pulling them up in 10 years because they look so dated? Slate comes into this category. It still has not come back into fashion and anywhere with slate tiles or pavers is considered very dated, even if they are still in great condition.

The more simple your choice, the more the pavers are the supporting cast not the hero of your garden, and the less likely they are to go out of fashion. Some materials, like sandstone, never go out of fashion, but supply is limited and the cost is high.

There are much cheaper concrete look-a-likes. They never quite look the same and it is questionable as to whether they are a better eco-choice. On one hand they both involve mining, though on the other hand they can last forever, which means they're not such a bad option from an environmental point of view. They can be much harder to match down the track if you wish to extend your paving, or if a section is damaged and needs replacing.

At this point you are likely to end up replacing the whole lot, even if they don't all need it. I have seen some quite fabulous paved areas created out of a collection of completely mismatched old pavers, and I've seen some very creative designs created out of two or three different pavers combined when there was not enough of either for the job.

The majority of us do not have the patience or the imagination to create something like this, so are better off trying to go with a paver that is easy to match. I have done small areas of paving with a mismatch of old pavers and bricks, and while I am very happy with the result, it is not a showcase result.

Hard surfaces direct water away from the local environment and into the stormwater drains. As this water is directed away the soil below the hard surface dries out and local drought conditions are created. A healthy tree needs to make use of the soil space below the hard surface but cannot do this if the soil is parched dry. By sloping the hard surface towards the garden bed slightly and adding a gravel trench, water is directed into the gravel trench where it can soak more deeply into the soil without flooding the surface of the bed. This allows for deep watering of the tree, and better growing conditions in the garden bed.

Garden edging

Garden edging can be created with a huge variety of materials and again, fashion can play a part. Rocks almost always look great. Bush rocks should never be actual bush rocks. Leave the bush intact and buy rocks from a landscape yard, in which case they are most likely to have been locally quarried. Even better, collect them over time as someone else is getting rid of them and advertising them free or cheaply online.

Old pavers can make wonderful edging. They can be laid flat to give a wide edge or dug in standing up for a slightly raised edge. This can be a good way to use up old pavers that are no longer what you want elsewhere, or for a random collection you have acquired over time. Old or leftover bricks can be used the same way. I recall as a child living in a new development, we would scour the rubbish heaps of new home sites for left over or broken bricks to bring home for mum to use as garden edging. It stuck with me and I have found myself doing the same thing more than once. These days we have a notion of 'old bricks' as opposed to new bricks. Age brings character, and you can even pay a lot of money for an old-fashioned style of brick. In a garden, let your new bricks get dirty, it will help them hold moisture and develop an aged patina.

I have a client who managed to get hold of a load of old-fashioned brick with loads of character. These bricks had large-ish holes in them. He laid them upright as garden edging, which not only showed off the character of the brick, but also created instant refuges for lizards and other small critters in his garden.

Besser blocks also make wonderful garden edging, although they do tend to look rather drab. A coat of paint can go a long way to brighten them up, or even make them completely disappear into the garden depending on the colour you use, which may also depend on what colour left over paint you can get your hands on.

Besser blocks are great for making raised beds and can be stacked a couple of blocks high. The holes can be filled with dirt and be planted up. This can be especially useful for planting things you do not want spreading into the rest of the garden. Besser blocks can be purchased new and in quantity, but are very widely used in building these days so are not hard to find as small amounts of unwanted left overs.

Timber sleepers or logs can also make beautiful garden edging, keeping in mind that timber will rot so take care with what you choose and the expected lifespan.

The advantage of any of these sort of garden edges is that they can be pulled out and moved or rearranged as the shape of the garden evolves over time. This of course cannot be done if you cement them in. Please avoid cementing in garden edges as much as possible. Not only does it mean that materials cannot be adjusted or reused later on, but it also adds unnecessarily to the energy and material inputs in your garden.

No matter how wonderful your garden design is, or which celebrity created it for you, gardens do not stay the same. Over time things will grow, things will die, neighbours will build a new house that cuts your light, or cut down trees which give you shade. Babies come and lawn is needed. They grow up and leave home and the garden can enlarge. Garden edges should be able to be moved and adjusted as the garden evolves.

Plastic garden edging is common and easy to use. Some of it is even made using recycled plastic, which should be considered an advantage over the ones that aren't. Plastic garden edging is very practical but rarely looks good. It is easy to use if you lay it in the sun to soften before trying to install it. It can easily be lifted and moved. It is also often lifted and thrown out because of its lack of visual appeal, so is often easy to get for nothing online or on kerbside collections. It is tough stuff with a much longer lifespan than anyone's appreciation for it, so reusing an old piece is a great way to get things underway in the garden.

These days there are also some very fabulous metal garden edging options. These should be seen as quite permanent. They have high energy inputs to create them, and once they are bent to shape, changing that shape is often not possible. If you should find yourself removing unwanted metal garden edging, do offer it to others. It is popular and can look fantastic, so can be used very stylishly in a garden. It is also made to last, so if you don't want it, pass it on to live a longer life in someone else's garden.

Metal is also a popular choice for raised garden beds. These preformed raised garden beds have become so enormously popular that people often buy them because they want a vegetable garden and think they need to do it in a raised corrugated bed. I have been amazed at the number of consultations I have done where people have bought the raised bed and got me to tell them where to put it and how to plant it. Most of the time, they have perfectly good garden beds on the ground that will be much easier to grow vegetables in.

Raised gardens

Raised gardens are popular these days for some good reasons. The main purpose of a raised garden is to reduce bending. This is particularly important as we age or for people with disabilities. Everyone should be able to garden, so if a raised garden increases accessibility, it should be embraced.

The other reason for a raised garden is to avoid soil problems. You may have contaminated soil, extreme soil conditions such as coastal sand or a rock shelf, or you may be trying to garden on a concrete slab. Basically, in all of these cases you are doing a version of container gardening.

If you have average to decent soil, and are able to bend, think twice about creating a raised garden. While you may have already guessed that a raised garden will require additional resources to build and fill, that is not the only reason to think it through.

Raised gardens can have challenges to care for. They tend to dry out more quickly and therefore need more watering than a similar garden in the ground. Nutrients are more likely to leach out, and therefore need more topping up. The soil in a raised bed also is likely to be significantly hotter (or colder) than the soil in the ground nearby, especially if you are using a corrugated metal bed placed in the sun. All of these factors mean that extra effort is required to garden successfully in raised gardens. There is a reason why you see so many neglected raised beds, or beds being offered for free online.

It is almost always cheaper and more successful to improve the soil you have than it is to bring in new soil. A garden bed in the ground will almost always be easier to care for than a garden in a container, even a very large container.

If you do still really want a raised garden bed, it might be worth looking into a wicking garden. A wicking bed is basically a means of turning your very large container garden into a very large self-watering container. It will help you to maintain your raised garden unless you are like me in which case you will need a reminder to water it, even if that is every week instead of every day. There are loads of different designs for creating your own wicking beds available online, and they can be adapted to almost any sort of raised garden set up.

As we have said previously, having success when you garden is a critical part of the process. There is absolutely no point spending money, energy and resources on a raised garden that might be doomed before it starts.

Herbicides and Weeds

Dealing with weeds in the garden is another area that often involves a lot of nasty chemicals. I am amazed at how many gardeners still have a bottle of weed killer in the garden shed. It is also another area of the garden where we are constantly told to wage war on the enemy. Supposedly, weeds are so bad that excessive use of poisons can be justified. As we are finding throughout this book, there is always another way to look at the situation, and there is always an alternative response. Just like so many other problems in the garden, when we look at it in more detail, the problem is leading us to a healthy solution, if we would just pay attention. When it comes to weeds, I am almost always asked "where did they come from?". The question we should really ask first in regard to the weeds in your garden is "why are they growing in my garden?". Weed seeds are everywhere. They blow in on the wind, are carried in on our shoes, and delivered fertilised in bird poo. Given that so many different weeds seeds find a way into our gardens, with and without our help, why is it that we are not being overrun with weeds?

Weeds grow where the conditions suit them best, as opposed to our garden plants which grow (or try to at least) where we put them. When their seeds land in a suitable place they will grow. If the seed lands somewhere unsuited, they won't grow, or at least not well. If you have a happy crop of weeds, you have conditions that suit them. Change those conditions to suit your garden plants instead of the weeds, and you will have less weeds without having to do any weeding at all. This may be a good long-term solution for the garden, but it does involve some effort. We all love a quick fix. A bottle of weedkiller is a very convenient quick fix, but it has drawbacks.

Non-organic weedkillers are highly toxic to soil biota. This means that the more you use them, the more damage you to do to your soil. The soil is the most critical element of the garden, so anything that damages your soil is only going to make gardening harder. Weeds generally are able to exploit poor soil conditions far better than our fussy garden plants can, so the more damage you do to your soil with weedkillers, the more likely it is that weeds will grow better than anything else. Which of course suits the manufacturers, because as the weeds thrive, you buy more weedkiller. Given that we are pouring insane quantities of toxic weedkiller into our environment every year, why is it that we are making next to no headway with the war on weeds? The more we use, the more weeds we have. This is not a deliberate marketing ploy on the part of the chemical companies, it is the once unrealised side effect of soil damage. This side effect is no longer unrealised, but it is not highlighted. I have seen far more successful long term weed control through good old fashioned hand weeding, then mulching the bed. No chemicals needed at all, not even organic ones. It is time to break this negative spiral, and this unnecessary reliance on chemicals. We can do this by understanding why we have weeds at all and combating the problem at its core.

Weeds are colonisers. That means they will come up in bare spaces. The more bare spaces you have, the more weeds you will have. The solution to this is not only easy, it is very rewarding. Grow more plants! If you fill your bare spaces with plants, there is just no room left for the weeds. If you weed a garden, do not leave the space bare. The one thing you can be sure of is that something will grow in that bare space. If it is not something you have planted, it is likely to be a weed. If you are not ready to plant in that space, consider saving your energy and leave the weeds be until you are ready, or at least mulch the bed to protect the soil and stop weed seeds germinating there.

Weeds can be great indicators of soil conditions. Some weeds like heavy compacted soil. These weeds usually have a deep tap root which makes them hard to pull out. These tap roots are capable of pushing through the heavy soil, where others cannot, and in doing so are playing a role in opening up the compacted soils. Many lawn weeds fall into this category, which is not entirely surprising when you consider that soil compaction can be a common problem in lawns.

Other weeds have very fibrous root systems which play a role in gripping into very sandy or silty soil, the sort of soil many plants struggle to get a hold in. By learning a little bit about the weeds you have, you will also learn a lot about your soil. Going back to basic soil improvement methods will help you to create a garden that is less favourable to the weeds, and at the same time better suited to the plants you actually want to grow.

The weeds themselves can be part of your soil improvement plan. Compost them to take advantage of the minerals they contain and put all this goodness back into your soil. Many weeds are highly adept at extracting minerals from the soil which are otherwise unavailable to other plants. When these weeds die and compost, those minerals are released in a form which other plants can now use. Weeds are actually a large part of natural soil improvement processes. If we leave them alone, they will proliferate, but as they do, they slowly work to improve the soil. In time the soil becomes more suitable for other plants and the weeds no longer dominate. This is part of the natural role of coloniser plants, not only do they fill empty spaces, they work to gradually make those spaces more suitable for more long-term plants to grow in. In nature this all takes time. In the garden we can speed up this process by appreciating this natural role of weeds. If you can find out what soil condition your weed is working to improve you can help out. Even if you don't know what the weed is and what it is telling you about the soil, you can be fairly confident that adding compost and doing some soil improvement will help your garden plants gain the upper hand on the weeds.

Weeds with seeds can be composted as we know that compost should not be left exposed on the surface of the soil but covered with mulch. Most small seeds will not grow if they are covered with mulch, for this reason mulch will play an important part in your weed management program. I personally prefer to compost weeds with seeds directly on the soil where I have pulled them out. Once a weed has gone to seed, there will already be some seed dropped in the soil there. I will have to go back and pull these out as they grow. By composting the weeds in the same place, I have only one patch to keep an eye on for weed seeds sprouting. By moving the weeds to the compost or bin, I have risked spreading the seeds.

Hand pulling of weeds is not always feasible. Happily, there are plenty of other ways to deal with weeds organically. The best method for you will depend on your situation and the weed in question.

Annual weeds, which reproduce via seed, are amongst the easiest to deal with. By stopping them from setting seed you can break the cycle and reduce next year's weeds. By far the easiest method is to cut them off at ground level and then mulch thickly over the top. The mulch not only stops any seeds growing, it prevents the plant from growing back.

This method is called smothering. It will work for any sort of weed that is not capable of growing back from an underground tuber, bulb or tap root. It will work on grass and other weeds with runners, however you will have to make sure there are no sneaky runners coming out from your smother layer. This method works via the depravation of light, which stops the plant from photosynthesising. If a runner is able to get out to the light, it will feed the roots under the smother layer and your efforts will fail. The mulch does need to be thick enough that the weed you are smothering is not able to force new growth through it. For this reason it is best for annual weeds, and is not so great for perennial or shrubby weeds.

For tough grasses like couch or kikuyu you may need a layer of cardboard or newspapers under the mulch to help hold them down. This method can be put to good use in creating new garden beds over old lawn. If you add a layer of compost above the cardboard and under the mulch, you are on your way to a no-dig garden.

You can up the ante on smothering by using a sheet of black plastic over the patch of weeds. This method is known as solarisation. It relies on a combination of heat and lack of light to kill anything growing below the sheet of plastic. Over the stinking hot summers we are so familiar with here in Australia, this can very effectively kill even the most tenacious weeds within weeks. When you pull off the black plastic, you may see white shoots. These are very delicate and searching for light. The sudden harsh sun will scorch and kill them. Leave the plastic open for a couple of days. If there is any sign of anything growing, cover it over again for a little longer. If you wish to use this method, think about the plastic you are using. Weed mat or shade cloth won't work as they are porous and will not hold the heat in. Ideally get yourself a sheet of heavy-duty black plastic that will not deteriorate in the sun too quickly and can be used numerous times.

Old carpet can be used for smothering but will not heat up sufficiently for solarisation. If using old carpet do remove it when the job is done. It will not break down and will be a permanent barrier in your garden. I have a client who lays an old rug over weed patches in a gravel path. The rug only stays there long enough to kill that patch and is gone again.

Weeds in gravel or the cracks between pavers are a key driver in sales of weedkiller. You can even buy formulations of weedkiller in ready to spray bottles marketed as path weeders. It is of course the same as the bottle just labelled weed killer, but this highlights just how much the weeds in pathways upset us. This is often the first bottle of weedkiller new gardeners will buy.

There are plenty of easy ways to deal with weeds in paths organically. Neat white vinegar is a very effective weedkiller and safe to boot. It is acidic and when sprayed on leaves will burn them. This will kill any plant that is not capable of growing back from the below ground parts.

If you have weeds with waxy leaves the vinegar should be diluted 50/50 with water and a dash of liquid soap added to help it stick to the leaves. Always rinse your spray bottle after using the vinegar. The vinegar is not toxic but can damage a cheap plastic spray bottle so rinsing will extend the life of your spray bottle. There is no need to add salt to this spray as is often suggested. Salt has an entirely different burning action and can cause harm in the soil. The vinegar will not cause harm in the soil.

A kettle of boiling water poured over weeds in a path is very effective. Boiling water is also good for weeds like nutgrass, oxalis and other weeds with underground bulbs, provided those bulbs are close to the surface enough for the water to still be very hot when it reaches them.

An alternative to boiling water is steam. Using your steam mop on your gravel or pavers will kill the weeds, including many seeds, thereby stopping the next generation of weeds. Steam is a very promising option for local councils to take up for managing weeds in public spaces. It leaves no toxic residue, kills seeds as well as weeds, and as the effect is instant there is no doubt about whether a patch of weeds has been sprayed with poison or not. The equipment set up costs are high compared to chemical systems, but ongoing costs are negligible, so it is surprising that it has not been taken up more by now. I suspect this is another case of out of sight out of mind, and without public pressure, there is little incentive to do things differently.

There is a growing trend towards wild foraging for edible weeds. This poses issues in terms of pollution. In urban or developed areas, the potential for weeds to have drawn up pollution from contaminated soil is real. Many urban parks are built over old land fill sites and the potential for toxic leachates is unknown. There is also no way of knowing if the weeds have recently been sprayed with weedkillers. Organic acid-based weedkillers (including cheap white vinegar) burn the plant and their effects are highly visible. Systemic herbicides such as glyphosate take many days to show any ill effects to the plant. They work by being absorbed by the plant, so washing the leaves will not make these weeds safe to eat.

Eat? Yes, many of our most common weeds are excellent and highly nutritious food plants. Chances are some of the 'weeds' you are trying to get out of your vegetable patch are better eating than your veg. Things like dandelion, flatweed, common mallows, chickweed, plantain, cobblers pegs, fat hen and soursob are all edible. What could be more sustainable than eating what is happily growing wild for you?

This is not just about taking advantage of what is freely available to you. Growing vegetables is very high input gardening. It is where a lot of people start when they turn to gardening, and when it goes wrong, it is where they leave their gardening dreams in tatters. (Of course, when it goes right, you have created a gardener for life, and they will eat well.)

Growing weeds on the other hand, requires pretty much no inputs at all. With a little care they will be lusher and tastier, but even with complete neglect they will provide you with a feed. Weeds feature highly amongst the edibles I harvest from my garden each day. I love that we get such a high nutrition output from something that did not need feeding, watering or fussing over.

I feel great about eating my weeds, but I have the added benefit of knowing that while these volunteer plants are feeding me, they are also feeding the insects, and feeding my soil. Weeds are agents of soil repair. They will slowly act to accumulate nutrients and improve poor soils. They are part of the problem imposed by humankind on the world and they are also part of the solution, as they slowly work to repair the damage we have done to soils globally. For this reason, if they are not in your way, or you are not ready to put anything else in that patch yet, leave the weeds be and let them keep working for you.

This same logic applies to lawns. Weeds in lawns are often the result of poor lawn care, which I will discuss in the next chapter. Lawn weeds are often deep-rooted weeds which are working to slowly bring nutrients to the soil surface and to open compacted soils. Instead of fighting them with chemicals that further damage your soil, why not help them with some soil care work? When people tell me that their lawn is a failure and is just weeds, I remind them that it may not be a great lawn, but it might just be a great meadow. A slight shift in our thinking can turn a problem into a desired outcome.

A meadow is characterised by a mixture of grasses and wildflowers. Does this sound like your lawn? Meadows are coming back into fashion, not just for their beauty but also for their vital role in supporting bees and other insect biodiversity. We have romantic notions of when a wildflower is good and when it is a weed. This often has us lusting after things that don't suit our climate, such as poppies in the subtropics. For a flower to be successful in a meadow, it needs to be able to cope with neglect and to reseed and come up again each year. In essence, it needs to behave like a weed.

What's more, weeds are often very pretty, once we stop choosing to see them as a problem. We forget that the flowers we see in the meadows from other climates are often a weed there. Perhaps take some time to notice who is visiting your wildflower weeds. Are they attracting bees and butterflies to your garden? If so, they are playing an important role in local biodiversity, and are contributing to a successful meadow.

To delve deeper into understanding and managing your weeds, I highly recommend you get hold of a great book called 'Working with Weeds'.

Sustainable Lawns and Hedges

A visit to any lawn website, or hardware store will reveal a plethora of lawn care products. Fertiliser, poison (even combined as weed 'n' feed), and lawn grub killer being the most common. Do we really need them? NO. Simple. With good care of the lawn and good sustainable soil care none of these are really needed.

There was a time when the idea of a sustainable lawn was an oxymoron. Traditionally lawns require high inputs of fertiliser, water, chemical treatments and petrol-powered care regimes. They tend to be a monoculture, that is, only one species. At least, when people are happy with their weed-free lawns they are a monoculture. All of this seems to fly in the face of sustainability. But there is more to the lawn story. Lawns are living surfaces, capable of absorbing rainwater, sequestering carbon and they do not absorb or reflect heat. The alternative to lawn is often paving or some form of hard surface. While lawns may not be as eco-friendly as diverse garden or meadow space, they are certainly more eco-friendly than concrete.

Being a living green surface, lawns can provide some degree of cooling when compared to hard surfaces. At least they can when they are green. Transpiration of water from the lawn cools the air above the grass. The cooling effect of lawns diminishes as they brown off during drought.

We are now recognising that being living green plants, lawns are able to sequester carbon. As we have already discussed, all living green plants will remove carbon from the atmosphere to some degree. Grasses have been found to be quite good at absorbing atmospheric carbon and storing it in the soil, when they are growing strongly. This includes lawns. The catch here is that a half dead lawn will not only not have much of a cooling effect, but it will also not be doing anything to store atmospheric carbon on the soil. A lush green lawn requires high water and fertiliser inputs. Clearly a trade-off needs to be made to find the sweet spot from a sustainability point of view. Happily, grass is tougher than we give it credit for and the sweet spot is not too hard to achieve.

If we let lawns dry out, they can turn brown and crispy very quickly. They will also recover very quickly when rain does come, even if they have been dry and crispy for months on end. In this state they are not sequestering carbon, nor are they providing any cooling, but they are also not using any resources so are effectively dormant. If you have water available to water a lawn, do so sensibly. As with all plants, lawns do far better with a deep watering less often rather than regular light sprinkles. Lawns too, need to be encouraged to push their roots deep into the soil, away from the surface where they will dry out quickly. A healthy lawn should not need watering more than once a week, if that, even in drought. Factors which affect a lawn's water needs (apart from weather) are fertilising, mowing and turf variety.

Fertilising lawns

Fertilising any plant will increase its water needs. Lawn fertiliser tends to be pretty strong stuff. It is usually around 20% nitrogen. At that level nitrogen can burn. It is not uncommon to see black patches in a newly fertilised lawn where it has done just that. Watering the fertiliser in well reduces the risk of burning, however as nitrogen is also highly soluble, watering in well will help reduce burn by washing away the problem. Regardless, plants do need more water in order to process the large amounts of nitrogen. Cutting back on the nitrogen will cut back on the amount of water your lawn needs.

A healthy and lush lawn can absolutely be achieved without fertiliser. You do this by caring for the soil. It may not be ideal to pile compost on the lawn, but you can sprinkle it on for great results. And if you have a great rock mineral product which is high in silica and soil microbes, sprinkle it on as well and stand back. This will help to reverse the compaction and enhance the natural nitrogen cycle to create a lush green lawn which is not only more drought tolerant, it will also be more wear resistant. Mowing without a catcher so that the clippings are returned to the soil will also help to close the loop. The more clippings you take away from your lawn, the more soil care you will need to replace that goodness. Obviously, this only works if you mow often and have only small amounts of clippings each time. Large clumps of clippings will block the light and give you bare patches.

If all of this fails and you do still feel the need to fertilise your lawn, swap out the lawn fertiliser for an organic fertiliser pellet, or fertiliser product that is less than 10% nitrogen, even better if it is less than 5%. Fertilise only when the lawn is actively growing, as a plant which is dormant or stressed by drought cannot absorb fertiliser.

Mowing

Mowing too short is probably the greatest crime against happy lawns. It will stress the lawn and can leave brown patches called scalping. These areas dry out even further, becoming more compacted and creating a perfect opportunity for weeds to enter. A lush thick lawn will have far less weeds as there is no space for them to get hold. Mowing is also a way to impact on the water needs of your grass. Mowing it short does NOT mean you get to mow it less often. It does mean the lawn will need more water to help it grow back and stay lush. A lawn, a hedge or a good haircut look and perform best if you cut a small amount often. Mowing it short and waiting for it to get long before cutting again is a great way to destroy your lawn. Grass should still be completely green after it is cut because the green part is the leaf. The grass plant needs to photosynthesize and to do that it needs green leaves. The more green leaf matter you remove, the harder it is for the grass to photosynthesize and feed itself. If the plant cannot feed itself, it needs extra help from you in the form of extra fertiliser and water. Leave a little more green on the lawn and it will better care for itself.

Turf variety

The other aspect we mentioned was turf variety. This one mainly applies if you are laying a new lawn and purchasing turf. Choose carefully. The more shade tolerant the turf is, the more water it needs, especially if laid in full sun. A finer-leafed grass variety will need less water than a broad-leafed grass. The native Australian turf variety is very well suited to our soil and climate conditions and is the most drought tolerant (e.g., has the lowest water needs) of the currently available turf varieties. Choosing the right turf for your spot will help to cut down on the level of care it needs. Planting a fine-leafed turf such as couch into a shady area will not work no matter how much fertiliser and water you give it. It is a full sun variety. As with so many aspects of sustainability, best fit for purpose goes a long way to cutting down wasteful excesses.

By choosing the correct lawn type for your situation you will be able to cut down on the amount of mowing needed. Zoysia turfs are gaining popularity for this reason. They are reasonably hard wearing, drought and disease tolerant and slow growing therefore need significantly less mowing than many other grass varieties. A zoysia turf can need half as much mowing as a kikuyu turf variety under similar conditions.

Turf farms are not well known for their environmental credentials. Apart from water and chemical inputs, a large problem with turf farms is that they remove a layer of soil every time turf is cut and sold. Soil being such a precious resource, this should not be considered lightly. Often turf is used because we don't know what else to do with a lawn in poor condition. I have created a beautiful lawn using runners that required weeding out of another garden. The grass had become a weed where lawn edges were not maintained, but planted elsewhere it made a glorious new lawn in one summer's growing season. A lawn in poor condition can almost always be restored by focusing on soil care and mowing a little higher.

Sometimes turf is the only way to go. Choose a type of grass that will perform well in your garden. Factor in the amount of sun the area gets, the amount of wear and the climate, then ask your local turf farm for advice of the best grass type for you, or research it online. It is better to set yourself up for success at the start given that turf comes at a significant cost to the environment, and your budget.

Lawn pests

Unfortunately, a number of pests are attracted to lawns. Top of this list are things like lawn grub, army worm and cut worm. At the first sight of small brown patches in the lawn, so many gardeners are running for the insecticides, but as with so many things, understand the problem first and the solution can be better targeted.

Lawn grubs are the larvae of a variety of beetles. This can include the larvae of some of our very iconic beetles such as rhinoceros and Christmas beetles. So many of us are lamenting the days of old when summer meant the night time buzzing of colourful and shiny Christmas beetles. It's pretty rare to see them these days, especially the fabulous metallic ones. Back then we didn't worry about lawn grub and didn't poison the grubs in our lawn. As our pride in lawns grew, so did the demise of the Christmas beetle. Beetle larvae (those curled up white grubs) eat dead organic matter in soils. They are a natural part of the compost cycle. If there is no dead organic matter in the soil, they will eat living organic matter, which means plant roots. They are a problem in lawns and pot plants; places with limited organic matter for them to eat. Part of the solution is to have well mulched garden beds nearby to create a preferred place for the beetles to lay their eggs.

Army worms and cut worms are both caterpillars, which will become moths. They hide at soil level and are hard to see, which is why they are referred to as worms, which they are not. They both feed on the green parts of the grass, are voracious feeders and can strip a lawn bare when present in large numbers. By the time we realise they are there it is often too late to do much about them. Your lawn has been stripped and they have already moved on to the neighbour's lawn.

All lawn pests love soft, sweet nitrogen fueled growth. The more you pamper your lawn with fertiliser and water, the weaker the plants will be and the softer and easier for pests to eat. Your pampered lawn is to pests as white bread is to kids – less challenging to eat and therefore easier to eat more of as it is less filling. If you are having trouble with pests in your lawn, put away the fertiliser and work on the soil to create a tougher lawn.

You may also find that by encouraging the local magpie family to visit you have less grubs in your lawn. A bird poking around in your lawn is a sure sign you have grubs of some sort. Let the birds feast on those grubs and save you having to treat them. If you have been poisoning lawn grubs, remember that poisoned grubs can mean poisoned birds so don't let the birds poke around this time, but next time, leave the poison in the shop.

If you do have a suspicious brown patch that seems be growing, start by pouring a bucket of very soapy water on it. This will kill most things in the soil, including lawn grubs, but is not as long lasting or as expensive as poisons. If it is just one patch, the soil life will be quickly restored from nearby as the soapy water breaks down. You can also try laying damp cardboard or towels on the lawn overnight. The grubs will come to the surface. In the morning, remove the cardboard and let the birds feast.

Lawn weeds

If you have a lawn full of weeds rejoice. Yes, leave the weed and feed alone and have a look at the message the weeds are giving you. They are telling you that the lawn is stressed. If they are largely flat leaved weeds like dandelions or plantain, you are mowing too short and giving these weeds the advantage. If they are deep-rooted weeds like creeping indigo, you have compacted soil which is making it hard for the grass to get its roots into.

If you have nitrogen-fixing weeds like clover, you probably need to feed the lawn – which you can do very effectively by mowing in the clover. 'Mowing in' simply means mowing without the catcher, so the clippings can fall back on the lawn to compost in situ. This is only effective if you are only cutting a small amount off the top of the grass. If the grass is long, there will be too many clippings to be able to leave them to compost on the lawn. If after mowing you have clumps or piles of clippings over the lawn, you will need to take them off or they will cause the grass to yellow and die back underneath them.

Enjoy the weeds in the lawn! They add character and produce flowers, which attract bees and butterflies. What would childhood be without dandelion wishes? There are a lot of ways to manage weeds in lawns without resorting to poisons. Never use a product that promises to fertilise your lawn and kill weeds in one go. It makes no sense. Why fertilise the same plants you are trying to kill? Yes, the two elements do somewhat counteract each other and reduce the effectiveness.

The key to having a lawn that is free of weeds is in the way that you care for the lawn. A thick lush lawn will not have weeds as there are no empty spaces for the weeds to take hold. Even if your lawn looks more like a meadow than a lawn, it can be restored with good care. Mow regularly and a little higher. Let the weed clippings compost back into the lawn. This degraded lawn is already so full of weed seeds that a few more will make no difference. The compost that the weed clippings add to the soil as they decompose will make a difference. Give the lawn a light feed with a sprinkle of compost and/or rock minerals, and a good watering. Then keep up the regular (fortnightly) mowing on a high setting.

It will probably take an entire summer, but gradually the lawn will be restored. You may have to dig out some of the more pernicious weeds, but don't do that until the grass is starting to dominate, otherwise the digging out only creates bare patches for more weeds to take hold. This is not an instant fix, but it is a very effective long-term fix as it creates good lawn care habits, and works on soil improvement.

Power tools

Having fewer hedges or lawns will go a long way towards reducing the use of power tools in the garden. Petrol-powered garden tools are noisy and pollute the air. They are, however, very powerful and often the most efficient way to get some of our bigger jobs done. It's useful to think about them in the same way we might think of using a car. If it is a small job, leave the tools in the shed and do it by hand. If you only have occasional need, call in the professionals. Choose an energy efficient model that suits your needs and you will also save money. Just like a car, it pays to keep your tools in good running condition to reduce emissions and extend their life. Good quality tools (of any kind) that do the job well are worth spending money on and caring for so that they will last. It can be very easy to spend a lot of money on tools, including power tools, that don't last. This ends up being not only a waste of money, but a waste of resources and landfill space.

There is conflicting information about just how bad petrol-powered lawn mowers really are for the environment. They are not as efficient as your car engine and as such will produce higher amounts of pollutants per litre of petrol than your car does. The exact figure has been reported differently so many times that it is impossible to know which might be accurate. A poorly maintained machine will run less efficiently and that applies to mowers as much as anything else.

It has been reported that in the US more fuel is spilt trying to fill the lawn mower each year than was spilt by the disastrous Exxon Valdez oil tanker spill in Alaska in 1989. We tend to forget that small amounts add up to an awful lot. Taking more care when filling the lawn mower and investing in a spill proof fuel can are great starting points for making a difference. Usually that fuel is spilt in our own garages or on our own footpath. This small amount that gets spilt regularly is all ending up polluting our own soil. For the sake of our own healthy gardens, it is a good idea to reduce fuel spillage!

Electric or battery tools will produce fewer emissions. Apply all of the above when choosing an electric tool. If it is not fit for purpose, it is more environmentally expensive than the petrol-powered alternative. Again, quality matters! Will you get the battery life you need? Be cautious about cheap battery-operated tools as the batteries often have a short life. A friend bought a lightweight battery blower from a fairly reputable brand. It was easy for her to use but the battery life was only about 10 minutes when fully charged, so it has sat and never been used again.

Is using power cords an option? I had an electric hedge trimmer but managed to cut the power cord so many times I reverted to battery. A decent battery powered hedge trimmer has done everything I need it to for many years now and with a bit of blade care is showing no signs of slowing down. If the job is too big for my battery hedge trimmer, it is going to be more than I can physically manage on my own anyway and it is time to call in help.

Using hand tools can save even more emissions: good old-fashioned push mowers, a spikey wheel on a pole to do edges, a broom and a pair of hand shears. These are much quieter, won't pollute and will save on gym fees. Again though, they need to be fit for purpose. There is no point buying these tools if you have a large lawn, unless you are a real fitness freak.

It is also worth investing in quality. Tools which blunt quickly will be very difficult to use and you will find yourself buying that heavy-duty petrol mower in no time out of frustration. Hand powered tools work much better if the job is not too challenging. If you have let your grass get knee high, no push mower is going to get through it. The longer the grass is, the harder it will be to mow. Small work-outs often will lead to successful hand mown lawns and great fitness. Of course, when you reduce the need to mow or trim a hedge, the use of any kind of tool is reduced.

Hedges

Take the time to plan a hedge before planting one. They can be wonderful garden features or an overgrown jungle that needs a lot of work to maintain. The concept of hedges originated in cold climates where they grow slowly and only needed trimming once a year. Hedges were featured in the gardens of the rich who had a small army of lowly paid gardeners whose job it was to keep them trimmed and neat.

These days the army of lowly paid workers is largely gone, replaced by one person and a powered hedge trimmer. The idea of the hedge is valued worldwide and has resulted in some very fast-growing plants used to create hedges in climates where plants grow fast. The faster your hedge grows, the more often it will need to be cut. Here in the subtropic we simply expect to cut hedges often. If you don't like cutting hedges often, think about how much you really need them in your garden.

There is another way. If we choose plants that only grow to the height we want the hedge to be, they will remain at that height if we prune them often or not. Trimming will help keep them neat but will not be the endless task it otherwise seems to be. This is far and away the most sustainable option for a hedge anywhere in the world. Choose your plants carefully. Match them to your climate. Check the mature size of the plant and don't plant something that wants to grow twice the size you want your hedge to be. A row of trees can rarely be restored to a hedge successfully, and yet it is something I am asked about often. There are so many cultivars of hedging shrubs (especially amongst lilly pillies and callistemons) available now that it is easier than ever to decide the height of your hedge, and then choose a cultivar which grows to just that height.

Chemicals – Pesticides and Poisons

Bugs and bees and birds

Lately we have been hearing a lot more about biodiversity in the garden and in particular the diversity of insect life that can and should exist in a healthy garden. As we rush through our gardens on the way to work, we barely notice insects. If we slow down a little, we may notice the odd bee or butterfly or grasshopper. As we take the time to stop and sit in the garden and really watch what is happening, we soon start to notice a variety of different bees, wasps, flies, dragonflies, beetles, bugs and other critters. I encourage you to seek out further information about the insects of your region, to notice what is living in your garden and see how over time as you grow more flowers, you will see a greater variety of insects. More and more good resources are becoming available these days as information about good bugs becomes more sought after, but there are still many insects that are barely known or not known at all. This is because they are, let's face it, small and not the most obvious things in the garden. A number of websites encourage you to record what you find and enter this information into their databases. These kinds of programs are not only informative for us, but help researchers gather information about the distribution of different species and even find new species. Gardeners are playing a very important role in furthering scientific knowledge of insects. Even gardeners in busy urban areas are finding new species of insects – and you thought the only species left to discover were in remote jungles!

All of this is actually extremely important to us as gardeners. Why? Because one of the key activities we all engage in is pest control. This is also one of the least environmentally friendly of the gardening tasks. Most of us have a shelf full of various poisons to help with the task. Even organic gardeners will find they have a variety of homemade or organic sprays with which they are armed and ready for garden warfare. In fact, there are many books dedicated just to organic remedies for various pests and diseases – entire books on how to kill bugs organically. Organic gardening has always been very focused on organic bug warfare, but even the organic sprays are killers. Sure, they are infinitely better than some of the powerful chemicals we have seen used in the past, but they are still general killers of whatever they come into contact with during their much shorter active life. Pyrethrum and nicotine sprays are examples. They are both derived from plants, and therefore organic, but they are also very good at killing insects of all persuasions.

Homemade white oil sprays are great, but they do suffocate all insects that they are sprayed on. There are times they may be beneficial in controlling a mealy bug infestation, for example, but use them carefully as you might just be killing the predators of the mealy bugs as well. A close look might be helpful here as one of the key predators of mealy bugs is the mealy bug destroyer (a type of lady beetle larvae), which looks very much like a slightly larger and more active mealy bug. This is not an uncommon strategy that insects use, especially if they are preying on insects that ants may be protecting. If you look like the critters you are hunting, you are more able to get close without anyone noticing.

To be a truly sustainable gardener, I encourage you to throw away almost all of your sprays. Instead, focus on healthy soil so that your plants are healthy enough to fight off the odd bug or two, and grow a diverse garden with lots of different flowers and plant types. The plant diversity is then providing food and habitat for a wide variety of bugs – good and bad, to make your garden home.

By allowing the good bugs to set up camp in your garden, they will do almost all of the pest control for you, saving you time and money on sprays, and saving the bees, butterflies and even up the food chain to birds and lizards from being wiped out by our poisons. In order for this to happen, we do need to leave alone some of the bad bugs too, as this is food for the good bugs. A constant supply of food means that there is a constant supply of good bugs feeding on them, and the system remains in healthy balance. If the bad guys are entirely wiped out, the good guys have no food, and will not be there when we want them to be. Be patient. You may need to hand squash an infestation that seems to be getting too much, but try not to spray, as the good guys are on their way. I know it's hard – I was getting very fed up with aphids on a client's roses. I had hand squashed them all twice and still they were back in their thousands on my next visit. As I pulled out the pyrethrum and started spraying, I noticed the ladybugs. Finally! I stopped spraying immediately and by the next visit, there were only a few aphids and the roses were flowering like mad. Balance had been restored.

If you have been using a lot of sprays or have not been growing a diversity of flowers, you may find it takes a little time, possibly up to a year for the good bug populations to build up sufficiently for you to be able to rely on them for your primary pest control. If you are struggling to be patient with this process happening naturally, there are now places that sell populations of good bugs which you can release in the garden to get things moving.

As I look out of my office window, I can see a mealy bug infestation at the top of my Rose of Sharon. It is only on a couple of branches, most are fine, and it is out of reach so I will leave it for whoever comes along to eat them. The tree is getting on a bit and they are short lived so this reminds me that it might be time to let a seedling or two come up and remain so that they can take over the place of the parent tree in a year or two. All the sprays in the world will not make this small tree live forever, nor will they give the tree the strength to soldier on. A bit of extra compost to the garden bed it is growing in and a good watering will get the best out of the last couple of years I have in this tree. As it approaches the end of its life and senesces it will be attacked more by bugs, which will in turn feed more good bugs. The branches and trunk will become softer, with hollow spots which will become nesting spots for lots of solitary bees and other good bugs and the leaf drop will add organic matter to the soil. And it has not given up the ghost yet – it is still an attractive small tree that flowers beautifully and provides welcome shelter for the garden below.

What are our pest sprays doing and are they really that bad? Well yes actually. The prolific use of organophosphates (and other similar chemicals) is now widely recognised as having pushed numerous species, including America's iconic bald eagle, to the brink of extinction. It was recognised that these chemicals are bio-cumulative, and in Australia they were banned in 1987. Bio-cumulative means that small amounts are stored in the fat cells in the body, not processed and eliminated. In this way every small 'non- toxic' dose remains in the body and gradually increases with each additional small dose until large and toxic amounts are stored within the body. This becomes a more serious problem the higher up the food chain you go, as each prey item may only contain small doses, but by the time a bird of prey or a dolphin eats many prey items they end up accumulating a significant dose of poison.

Since the ban on DDTs, we have turned to other poisons. Neonicotinoids (a synthetic chemical related to nicotine, and sadly as addictive to bees as nicotine is to us) has been heavily used. This substance has been linked to mass bee deaths worldwide, prompting a European ban on this chemical in 2013. At the time of writing these chemicals have yet to be banned in Australia, however public pressure has seen them removed from sale in hardware stores and nurseries. As this chemical can be used as a slow-release tablet placed into the soil, it has been popular with production nurseries for putting into pots being sent to nurseries. Where the shop lights are left on overnight, they attract moths which lay their eggs on the nursery plants. As the general public are not keen to buy plants chewed by caterpillars, these plants are all treated to prevent insect attack. The same public pressure has seen production nurseries move away from neonicotinoids placed in the soil of potted plants, but where one chemical loses favour, history has seen it replaced with another.

There is a possibility that as the European honey bee is feral in Australia, protecting it through banning harmful chemicals is not considered environmentally beneficial. This argument is pointless when you consider that (honey aside) almost all of our food crops are pollinated by the European honey bee, so without it we are likely to go rather hungry. Neonicotinoids harm all insects which come into contact with them, so while other insects may not be as economically valuable as the European honey bee, they are no less likely to be lost.

These are two extreme cases when it comes to garden chemicals. Many other chemicals may not be this damaging to our greater environment, but they are still having an impact. Most pesticides that we use, be that pyrethrum, white oil, neem, or even personal and household insect sprays are not specific, meaning they will kill whatever insects they come into contact with - good, bad and completely harmless. Organic caterpillar control sprays which contain bacteria specific to caterpillars will not harm other insects, but will kill ALL caterpillars so it may well save your cabbages but if used too liberally you will find yourself in a garden devoid of butterflies.

The simple answer to all of these problems is good bugs. Use pesticides sparingly and plant a variety of flowers to encourage good bugs into your garden. In most gardening books we are told to spray problem insects as soon as we see them. My advice is the opposite. Don't spray at all unless you find the problem is getting out of hand. I allow the odd bit of mealy bug, scale, grasshopper or aphids in the garden and only pull out the sprays when the infestation gets severe. This is rare, because usually as the pest population starts to build up, so to do the pest predators and the garden has a natural balance. While I wait for the good guys to do their job, I am not at all averse to squashing a few bad bugs as I find them, but I certainly don't have time to go around hunting them down.

There are always exceptions. One of these is cycads, in particular *Cycas revoluta* (sago palm). Since the introduction of the blue cycad butterfly a few years ago, they have been completely ravaged. Native macrozamia cycads are also attacked by the moth, although are more likely to be attacked if there are sago palms in the area. Let down your guard on the spraying regime and you will have lost an entire season of new fronds in a matter of days. I find a spray mix of a caterpillar specific product and neem oil works well but watch the spiders – they are doing their bit by catching the moths too. The other thing I like to do where I have a garden with cycads is to add in a few cardboard palms (Zamia species and actually cycads not palms). These will also attract the moths but are naturally toxic to them, killing the hatchling grubs as they start to feed and before any damage is done to the plant.

Another exception is gardenias. These are tropical and subtropical in origin so you would think they would be easy, and generally when they are included in a mixed planting they are. The problem tends to be greatly escalated when they are mass planted as they often are in modern landscaping or used in pots where they tend to dry out. Gardenias are beacons for scale and with the help of the most annoying pest of all – ants, they can become an unhappy weak mess of sooty mould very quickly. Regular applications of a homemade spray of one tablespoon of vegetable oil, one tablespoon of detergent in one litre of water will do the trick, but to have ongoing results you need to look at why the plant is being attacked. Keep the soil consistently moist (too wet or dry will only make the plants more unhappy) and give them a good feed with compost and rock minerals. Hose off the sooty mould, then a foliar spray of seaweed solution will help. Gardenias are acid lovers, so if they are doing badly, a pH test might show the problem to be in the soil pH. If all of this still doesn't work – and sometimes the combination of ants and scale is not that easy to beat, a spray with an organic horticultural oil is needed.

Yet another exception locally is agapanthus, another garden favourite which has succumbed to the mass-planting-equals-pest buffet equation. In this case, mealy bugs, and hiding right in the centre of the leaf shafts. The homemade oil and detergent spray mentioned above should work. One of the reasons why agapanthus are often badly pest affected here in the subtropics is because they are a cool climate plant growing outside their preferred range. Any plant growing in less than ideal circumstances will be far more prone to pest problems than if it is growing in its right environment.

For any plant that seems to be constantly bothered by pests, look into what else might be going on. Pests are naturally more attracted to weaker or stressed plants. The two major ways we create these weaker plants in gardens is by over fertilising, and growing plants unsuited to the location. We have already discussed the way that high nitrogen fertilisers increase a plant's need for water and produce soft sweet growth attractive to pests. Less nitrogen and a more balanced diet through substituting fertilisers for compost can address this issue. If you are still having pest problems, perhaps this plant is simply not suited to your garden. This could be (and often is!) a broad climate issue with cool climate plants being grown in hot climates and vice versa, or it can be a microclimate issue. Is this plant getting too little or too much sun? Is the soil too heavy for it? This takes us right back to the beginning of this book and understanding the conditions in your garden. You would be surprised how many times I am called in to help with garden problems which all stem back to plants in the wrong spot in the garden.

In my own garden, I very rarely notice any pest problems at all. If I do however find a plant that is being attacked, I first treat it with rock minerals. I use a product that is high in silica and therefore I can use it as one would use diatomaceous earth, also a silica product. I sprinkle it over the pest itself. The silica will desiccate the pest, and while it is unlikely to kill them all, it will help to reduce their numbers. Two days later I wash this down into the soil where it can help to improve it.

As the soil becomes more friable, more biologically active and more mineral rich, the plant will grow stronger and healthier. If this is not enough for this particular plant to recover and no longer be bothered by pests, I assume this plant is not suited to my garden and it can go to make way for something that is.

As mentioned, mass planting does most certainly lead to a pest buffet. Everything they like to eat all in one big spread, no hunting necessary. The message here is avoid mass planting. Mix things up as much as your sense of order will allow. This goes for the vege patch as well as the ornamental garden. In this way, when the pests find one of their favourites, they have not by default found them all, so some of the plants will remain unfound and untouched.

It also allows for companion plants to hide the pest's food of choice by perhaps masking the smell, or by distraction (e.g., a more preferred pest plant as a sacrificial alternative), or by instead attracting pest predators. A friend has found that by leaving a weedy castor oil plant in her garden (pruned to keep it manageable) the grasshoppers will eat it and leave her cordylines, which grasshoppers love, alone.

On the subject of grasshoppers, if you do not like them in your garden, do not knock down the spider webs. Although spiders rarely catch grasshoppers, studies have found where there are more spider webs, there are less grasshoppers. Grasshoppers lay their eggs in the soil. Mulch bare soil well so they cannot get to it to lay eggs for next year's pest problem. Grasshoppers also don't like shade, so having some part shade in the garden will reduce the attraction to grasshoppers. Robber flies, dragonflies, frogs and birds will all eat grasshoppers. A bird bath can actually be extremely useful for pest control. The birds that visit your garden will be more than happy to relieve you of a few pests while they are there.

All of this leads to a far more natural style of gardening, a style of gardening which emulates nature. It also leads to a garden without much, if any, need to carry out any pest control activities at all.

Fungal pests

That may have covered the insect pests but often times the problems gardeners face are not caused by insects at all, they are fungal or plant diseases. The answer to these problems lies in the overall health of the plant. A healthy plant is less likely to be attacked not only by insects, but by any disease problems. If you have problems, go back to the plant and look at why it may be stressed.

Sometimes fungal issues are more prevalent during hot wet weather. This may seem perfectly natural, but if you have climate suitable plants that are growing strong you should still not have significant problems. A little bit of powdery mildew is food for the yellow spotted lady bugs and usually is more of a problem on older leaves. If the whole plant is attacked you have a garden out of balance.

Many leaf fungal issues can be prevented in the soil. For every pathogen that attacks our plants, there are soil microbes that attack that pathogen. It is the good bug/bad bug story all over again on the micro level. A rich soil biota is critical in disease prevention in the garden. For this reason, always use an organic, microbially enriched fertiliser if you can - or add compost. Compost is critical for healthy soil biota.

Another way that the soil can impact on the spread of diseases to plant leaves is through splashing. As waterdrops splash onto bare soil they rebound onto the lower leaves of plants, carrying pathogens with them. Mulch prevents this happening, as does a densely planted garden.

I have a handful of roses in my garden. One of the key issues rose growers are told to look out for and treat is black spot, a fungal disease. Black spot is considered especially problematic in hot humid climates like mine, and yet I have never had any black spot whatsoever. My roses are grown in just the way we are told not to grow roses, crowded into a garden full of other plants. While this leads to increased humidity around the rose plant, it also means those other plants protect the rose leaves from any splash back from the soil. They are fed only with an occasional sprinkle of rock minerals and chop and drop composting. This has allowed a healthy balance of microbes in the soil which can keep fungal issues in check for me.

Some issues will still occur. I do get rust on my frangipanis. This has been a horticultural talking point in the last few years with a lot of worry about controlling a disease which has only been here for a few years now. We are told to spray the tree with a fungicide, then collect all the fallen leaves and bag and bin them. I don't do any of that. The rust only takes hold late in the season when the tree is about to lose its leaves anyway. At this point the tree has turned off its disease defences in the leaves as it withdraws nutrients from the leaf before dropping them.

All deciduous trees do this. The nutrients in their leaves are too valuable to waste so many are reabsorbed before dropping the leaves. It is this process which contributes to autumn leaf colour, as chlorophyll (which makes leaves green) is withdrawn from the leaf.

This process means that the rust is not attacking the tree, only the old leaves about to drop. The tree is finished with those leaves anyway. It causes no harm to the tree. Spraying fungicides may seem harmless until we remember that fungi are a critical part of a healthy soil biota. Like with everything else, there are good and bad fungi.

Amongst the good fungi are the mycelium which attach to plant roots, helping to feed the plant. We are aware that this is a highly specialised relationship that many native plants have but in actuality, most plants have some degree of fungal mycelium living in symbiosis with their root systems helping them thrive. This relationship is only beginning to be understood in detail but is certainly a very good reason to be cautious about using fungicide in the garden.

I also don't bag and bin the dead frangipani leaves when they have rust in them. I've noticed the rust helps the leaf litter to break down faster, which is after all a critical ecological role of fungi. Various bee species have been observed collecting rust fungi from plants. Studies showed that the fungus is a source of protein for bees. This can be very valuable for them when other food sources are scarce and they are preparing for winter.

Other fungal issues, things like phytophthora, are more serious. These can be highly persistent pathogens that can result in tree deaths and the soil being completely unsuitable for certain more vulnerable plants. Even with the use of fungicide these severe fungal pathogens are hard to control. Good gardening habits still play a big part. I once had to replace an entire lawn due to phytophthora. The pathogen had attacked a poinciana tree on the lawn. It was treated and the tree recovered, however the constant over feeding and watering of the lawn helped to harbour the fungus. This lawn was so heavily treated with so many different chemical treatments that there was no defence left by way of good soil microbes. The grass was weak due to the poor soil conditions, so was more heavily fertilised, producing soft growth which attracted pests. This entire situation became such a negative loop that the lawn had to be completely replaced. The new lawn was almost immediately badly overwatered and was attacked by the phytophthora. At this point not only did the lawn need to be replaced again, but so did the soil. This is a rare situation, but it highlights the negative spiral we can end up on by using chemicals to treat problems in the garden.

Much of this cycle of problems could have been avoided if the first step was to add compost and improve the microbial life in the soil. There are some very specialised products available these days which are essentially liquid treatments of soil microbes. If your problem is greater than compost alone, then I would encourage you to use these before turning to chemicals.

Often pest problems affect one group of plants particularly and can be avoided by removing this group of plants from your garden. This will not make the vegetable grower with root-knot nematodes in the soil happy. Root-knot nematodes attack members of the Solanaceae family, in particular tomatoes. We can try companion planting of marigolds and mustard, then digging these in, but ultimately the solution lies in soil microbes. The predator of root-knot nematodes is another type of nematode. The best way to have this present is to use a variety of sources of compost and to care for the microbial life in your soil.

If you are finding one group of plants are particularly affected by disease problems in your garden, you may have one of many different fungal, viral or bacteriological diseases persisting. Remove this group of plants and avoid growing anything closely related to them, as many plant diseases are specific to plant families. In these sorts of extreme circumstances treatment is very difficult. Avoid that group of plants for a couple of years while you work on improving the microbial health of your soil. Crop rotation in intensive edible gardens can help to break any disease cycles.

Plant Choices

There is no great art to choosing plants which are more likely to work in your garden, just a bit of extra thought and local knowledge. It is worth putting some effort into knowing what is likely to work. There is nothing more disheartening than a constant stream of dead plants. It will not give you any great confidence as a gardener.

Choosing plants that are going to die is not just a waste of money, it is a waste of all that has gone into growing those plants. We will all lose plants at some time, but hopefully we will learn from the losses and try not to repeat the mistakes. Hopefully, we will manage to keep more plants alive than not. Dead plants are a key incentive for people to engage me for garden consultations, so there is a risk the information I am about to share will put me out of work!

The greatest difficulty is in knowing which questions to ask, so this chapter is designed to help there. The key here is to ask questions. This usually starts with reading the plant labels. They are not always fabulously helpful, but it does surprise me how many people don't read them at all and therefore miss even the most basic information about that plant. The plant label will tell you some basic things like how big the plant will get and how much sun it likes. Pay attention to this! The first thing to get right is to purchase a plant that suits the space you have in terms of the size and the amount of sun.

The plant label will usually also tell you how much water the plant needs. It won't tell you which climate it prefers, so this is something to be aware of. A local nursery will usually sell plants best suited to the local conditions. A large chain store will sell whatever people will buy. It amazes me how much lavender gets sold here in the subtropics. Lavender hates humidity and dies readily here, but people love it and go back to buy more, so it is a good seller, if not a good survivor. This sort of thing is less likely to be an issue when shopping at smaller nurseries who will lose business if the plants you buy from them keep dying.

In these days of water restrictions and water shortages, the term 'drought-tolerant' is becoming something of a catchcry. A plant's water needs are now very often included on the nursery label, which is very handy information as we are being encouraged to purchase drought-tolerant plants. That is all well and good if you are experiencing drought, but what if you are not?

While there are many drought-tolerant plants that are quite happy to get lots of water, there are equally many that are not. I tend to be a little cautious when buying drought-tolerant plants, especially as I have heavy clay soil which tends to waterlog. I usually avoid the ones labelled as 'extremely drought-tolerant', and instead try the ones that are just 'drought-tolerant'. The reverse of this is also true. These very drought-tolerant plants will stand a better chance in my mother's garden 30 kilometres away as she has very sandy soils, although care still needs to be taken with regard to humidity. For her garden, plants with high water needs, such as heliconias, are best avoided even though they are well suited to a subtropical climate. This is why I have encouraged you to get to know the conditions that are affecting YOUR garden. It is a game of matching plants to the spaces you want to grow them in, only this is something of a 4D puzzle which includes not just height and width, but soil conditions, sunlight and local climate conditions.

Before I impulse buy a plant that I'm not familiar with (it happens often), I first try and find out as much as I can from the seller. If I know they have grown it locally, I know I'm in with a chance. Then I try and research the plant online with particular interest in where the plant originated. Anything coming from arid areas (cool or hot) is less likely to tolerate humidity. At this point we open a can of worms, because for everything that doesn't work, something else does. Your garden's soil type and local microclimate will play a big role in what works for you. Knowing a bit more about what this plant likes also allows you to match it to the right spot in your garden.

As gardeners we do need to take a risk and experiment a little. I have a fantastic collection of succulents. Most of them are in pots and growing better than those in the ground. I have found that many of the gorgeous rosette forming echeverias and the aeoniums never do well for me. They don't like the humidity. I have seen some stunning aeoniums around but they are usually grown by enthusiasts who are catering for their microclimate needs. I can still have glorious succulent displays, just using different species than I see in the pictures of temperate gardens. When I find ones that do well in my garden, I use them abundantly and try not to lust after the ones that don't.

Other drought-tolerant plants to avoid in the subtropics are most of the West Australian natives, many of the South African proteas, new releases of Australian arid zones plants, and grey leaved Mediterranean plants. None of these tolerate humid conditions and are even less tolerant of wet feet. You may succeed in growing them for a short while, but they will never be as wonderful as they are when grown in their preferred climate. Similarly, I have found poor success with many of the ornamental grasses originating in temperate climates. I adore the romantic pictures we see of miscanthus grasses swaying between salvias, lavender and grey cotyledons in southern gardens, but this is unrealistic in my part of the subtropics. In fact, many of the pictures we see of water wise gardens are more grey than green. In one delightful book in my library, there is only one subtropical garden featured out of 25 dry climate gardens. This is because we are not a typically dry climate, even though we may have decade-long droughts and severe water restrictions. Regardless of which broad climate zone you are gardening in, you will need to best match plants to that broad climate type, rather than what you may be experiencing in that particular season.

I may admire these temperate dry climate gardens, but here, I style them a little differently. A xeriscaped garden can still work very well in the subtropics if planned to allow for the wet times, however, I do feel that a lack of deep greens seems unnatural here. Allowing for the wet times usually means planting on a slope, preferably west facing, so that excess water can drain freely. It also means mulching with sand or gravel and allowing space between plants to reduce humidity build-up.

When making plant choices your style of garden will of course come into play. Even in sustainable gardening we are working towards creating a garden that is beautiful and satisfying to us as individuals. We may not be able to grow everything that we want, but we should still be able to grow the style of garden we love. Plant choices will depend on your local climate and your soil conditions – and then on water availability. This is because your local climate and your soil type will be a major influence on water availability. The other reason is that very broadly speaking, if a plant doesn't like your climate or your soil, no amount of supplementary watering will change that.

Plants growing wild

We can cut down the amount of work involved in caring for a hedge by choosing plants that grow to the height we want the hedge to be, and no higher. This concept extends beyond hedges into many other plants in the garden. So much garden maintenance, and so many garden mistakes all relate to plants growing bigger and further than a gardener wants them to. If before we plant something, we take a moment to understand the natural growth habit of that plant, we can set ourselves and the plant up for success. Constant pruning, shaping and manipulating plants means that we are not allowing them to grow in their natural form, and we are working against nature. This does not mean we can't give them an annual tidy up; it just means we are creating unnatural, high maintenance gardens. Vines are a great example of where we often go wrong with plants growing true to form.

Vines are adapted to grow in forests where light is in short supply. They aim high in search of light. By understanding this basic instinct of vines, you will have greater success with them. Don't plant them on trees, unless you want the canopy of the tree taken over by the vine. If you would like the vine to spread laterally to cover a fence or a wall, you will have to help it. A vine's natural instinct is to go up, not out. Without your help to constantly redirect the branches laterally, they will simply grow up and along the top of the fence, leaving the side of the fence bare. Note the word 'constantly'. Anything that is not natural will take more work to sustain. Only you can decide if the result is worth the effort. Part of being a garden ecologist is to respect the nature of the plant you are trying to grow and to work with it rather than against it.

I had a client with a wisteria growing over a patio. The wisteria had developed a wonderful trunk and dense coverage on the patio. It did not stop there. In our subtropical summers, it headed up the house on one side of the patio, the tree on the other and reached out and attached to the neighbour's house just over the fence. Every two or three weeks this wild wanderer needed to be pruned to keep it in check. The constant pruning meant constant work with power tools that could be eliminated if a less vigorous vine were chosen. It also meant the wisteria never flowered well as the flowering stems were constantly being cut off.

By contrast I have a variegated golden chalice vine growing on the front of my house. It is a slow growing, woody vine and fills the space beautifully. The flowers are dramatic and it is a significant talking point in my garden, in or out of flower. It needs pruning only once a year with hand tools. In over 20 years I have never needed to use power tools to control it.

The decisions to be made here relate to how much effort you are prepared to put in. If you want the exercise then great, but if that means more use of power tools and lost weekends, perhaps a more sustainable option could be investigated, one that allows a plant to grow to its true potential in the space you have.

When planting vines on a fence I encourage you to also think about your neighbours, especially if you are planting on the shadier southern side of the fence. I have seen jasmine, passionfruit and other vines reach the top of a fence and then travel into next door's garden. This is just plants doing what they do, growing and reaching for the sun. If we want our garden to be a successful ecosystem, it would be great to have neighbours who also embrace the ecology of both gardens, and this is not going to happen if they are constantly fighting your vines.

Climate appropriate plants

Gardeners have long held a fascination with plants that are exotic. While the word 'exotic' simply means 'not from around here', to us it is so much more than that. 'Unusual', 'hard to grow', and 'stunning' spring to mind. Most often we are looking at a plant from a different climate zone to the one we are living in. It may be pictures in a magazine, or it may be travel. A friend came back from France 'oohing and aaahing' about Monet's garden and the gorgeous flowers growing everywhere there but noticed that the people she met there were far more interested in 'oohing and aahing' about the things she had in her garden back in Queensland – strelitzias, heliconias, and bromeliads. Another friend came home from England disappointed that while she cannot grow the things they have in their gardens, they seem to have all the things she grows in hers here in the subtropics. They do have an advantage – conservatories.

Conservatories originated in Europe in the 16th century when they were primarily used to grow citrus from the Mediterranean in colder climates. In that form, they were known as orangeries. As large sheets of glass become more readily available, the concept of the conservatory developed. Technically a conservatory has more than 50% of its wall surface glazed, and these days modern technology goes so far as including not just double glazing, but also argon impregnated glass, heat reflective films and thermal ribbons. That is a lot of trouble to go to in order to grow something from a climate incompatible with the one you live in!

For us in the subtropics, the reverse is not so simple. We have heat and humidity. We tend not to have large well-lit spaces with lower temperatures and no humidity in which to grow the cold climate plants that we are fawning over in magazines. Indoor air-conditioning does not really work as the plants we are after need direct sunlight. Indoor plants are most often tropical understory plants which can tolerate low light situations. They don't like air-conditioning either as they do like humidity. For the truly dedicated or those with plenty of time, we can put bulbs in the fridge for six weeks a year to chill them, and we can put ice cubes onto the cymbidiums every evening for a month after Easter to give them the chill they need to set flower, but for most of us, this is not going to happen.

As gardeners, we all fall in love so easily with wonderful plants and flowers, it is impossible not to. Restricting ourselves to growing what works in our climate is not always easy. In these days of television, the Internet and aeroplanes, we can travel so easily (virtually or actually) to so many places and climates and see so much that inspires us and ignites our passion for plants. Plant breeding, world trade and migration have made so many plants more accessible to home gardeners regardless of climate suitability.

In a country like Australia or the USA, it is possible to move interstate and pack your favourite plants in the car, only to plant them in your new garden and watch them die because you have moved into a completely different climate. That being said I know of one elderly gardener who has a rose growing in Brisbane that is quite well travelled. Her grandmother brought a cutting with her when she travelled from England to Sydney in the late 1800s. Her mother took a cutting from Sydney to Cairns when she got married, and this lady brought a cutting from Cairns to Brisbane where it is now thriving. Great (true!) story but more often than not, it does not work out so well.

To be fair, climate zones are not always true to type, so knowing what grows in a climate zone and sticking to it is not a black and white guide to what works in your garden. I know I said this earlier, but I urge gardeners to know what their localised climate is before you start thinking about what will grow there. It is entirely possible that although you have moved into the subtropics, which technically does not go much below 6°C, you might be in a pocket that regularly gets frosts. Or you might be in a coastal position where the regular sea breeze keeps humidity a little lower, or be surrounded by a pocket of rainforest where the humidity is always a little higher.

Climate will dictate so much of what you can or cannot grow, so it is worth knowing just what your climate is. I love cottage gardening. A style of gardening that arose in Europe's cold and temperate zones, where summers are mild and winters very cold. It's pretty much the opposite of where I am trying to copy their style of gardening. I encourage you to be inspired by gardens everywhere, take their design ideas and apply them to your own garden. When it comes to the actual plants used, what substitutions can you make? If you are in a cold climate hankering for a tropical look, use plants with large leaves. Where palms aren't happy, try tree ferns. Gunnera or rhubarb will give a lush look with shade and water, and add in things like fatsia and hostas. Choose plants that flower in hot colours such as red and orange.

In warm climates substitute miscanthus grasses for native grasses such as kangaroo grass, barbed wire grass or native pennisetums. Interplant these with some macrophylla salvias (*Salvia gregii* is not good with humidity) and smaller leaved flowering perennials in the softer colour range of pinks, blues and purples. Salvia 'Anthony Parker' is a grey leaved salvia with purple flowers that truly thrives in the subtropics. Instead of heucheras try bedding or rhizomatous begonias.

When looking at a picture of a garden that you are inspired to copy, first find out where in the world that garden is. If it's not in a similar climate zone, focus on the style rather than the plants themselves. Look at things like the size and colour of the leaves, flower colour and size, overall shapes of the different plants, textures and plant forms, and how these relate to each other in the space. The design then gets broken down into something such as large green leaves on a tall plant paired with smaller red leaves on a medium height plant, with clumps of low growing small leaved plants with white flowers. This design can now work anywhere, by finding locally available and suitable plants that fit those descriptions.

While we may always dream of plants from another climate, attempting to grow them in our own climate is fraught with difficulties. Most often plants grown in unsuitable climates do not perform to their true potential and need a lot more care just to survive. I regularly mention that grey leaved plants do not like humidity. Someone recently pointed out *Plectranthus argentatus* to me. I have always thought of this plant as a very reliable grey leaved plant that thrived in the hot humid subtropics and so held it up as our go-to plant for grey leaves. That was until I saw it growing in a slightly cooler and less humid environment on nearby Mt Tamborine. What I thought was a very happy and successful plant in my garden, was not even close to reaching the full potential of that plant!

Tropical blueberries are another great example. The 'tropical' bit refers to the fact that they need less chill to set fruit than do ordinary blueberries. It does not mean they will thrive in the tropics. So many subtropical gardeners proudly show me their stunted, sparse little plants with a handful of fruit on. Anyone reading this from a cold climate will know that blueberries grow on a very large, lush bush, and would not recognise the little plants that are being nurtured in a too-hot climate.

It is often because we desperately want to grow a plant that we overlook its poor performance. Lavender is a great example here. It is widely grown in the subtropics, a climate very different to its natural Mediterranean climate. Lavender is a plant that seems to evoke much emotion, and gardeners everywhere are desperately keen to grow it. Some varieties (particularly French lavender) are purported to do better than others in humid climates, but the truth is that none really like the humidity, and it is very rare indeed to see lavender growing well and reaching its true glorious potential in humid climates. It is however very common to see it struggling along with more die back than flowers, but loved nevertheless.

Does it really matter if our plants are not reaching their full potential if they are bringing us joy? Not necessarily – go ahead and experiment and enjoy your garden. In fact, with a changing climate we may find that experimenting with plants from locations that are a few degrees different from our own is key to adapting to our gardens to climate change. Climate scientists recommend that we to hotter and drier climates by about one or two hundred kilometres to find plants which will suit our climate in the next 30 or so years. In Australia that means north and inland from where we are. If you are reading this in the Northern Hemisphere, you will be looking south to find the hotter climate zones.

However, if the experiment is not so successful, you will find yourself with high care plants. High care plants in most cases are not good choices for sustainable gardens. They tend to need more water and are often more susceptible to pests and diseases, resulting in more sprays, and a greater need to replace them as they die. These plants also tend to get more fertiliser as we attempt to help them along, even if extra fertiliser is not what they need at all.

Of course, climate is not the only appropriateness factor – plants planted in the wrong soil type, wrong light or water requirements will all languish and perform badly, becoming a magnet for pests and diseases in the garden. So often I am confronted with gardens that have suitable plants but they are struggling in the wrong position. A recent consultation had water-loving heliconias at the top of a slope which is the driest part. It had aloe (a sun loving succulent tolerant of dry conditions) planted with alocasias (a tropical plant needing shade and water) in a dip at the bottom. All of these plants will do well in this garden if they are rearranged to be planted in more suitable spots. If they are not moved, they will all suffer.

It should be noted here that if you have a plant that is suffering, it can harbour diseases that are hard to get rid of, even with the use of nasty chemicals. If you notice that a plant is looking rather diseased, it is best to get rid of it before the problem can spread to other plants. In the case of plant diseases, do not add them to the compost. Instead tie them into plastic bags and dispose of them in the rubbish (not green waste) to limit the spread of the disease.

These high maintenance plants therefore do not fit with the sustainable garden in which we have taken steps to reduce the use of sprays, fertilisers and water. Eventually we need to make the choice that these plants are not carrying their weight in our sustainable garden and will benefit us more by being added to the compost and making room for something more appropriate.

Enjoy the things that will grow well for you, and then enjoy admiring something very different when you travel. Afterall, travel would not be as enjoyable if you were to only see what you already have at home.

Sourcing plants

Something as simple as where we source our plants from can impact on the sustainability of our choices. The distance a product travels before it reaches us is something we often hear in terms of food and food miles, but it relates to all products we consume – including landscape supplies, gardening materials and plants.

Plants are living things that like certain climates. By sourcing plants locally, they are far more likely to have been grown in a similar climate and therefore do well in your climate. Plants from outside your climate may acclimatise and be ok, but many will languish and perform badly if they live at all. Experimenting with plants from different climates can be a lot of fun, but there are usually several heartbreaks for every success. Still, in these times of climate change, it could be well worth thinking about plants from a warmer or drier climate and see how they go for you. Generally, things that are not too far outside your climate zone will do much better than things from a completely different climate, and will not have as far to travel.

Many nurseries, especially the large hardware chains, stock similar plants in most of their nurseries. These plants are often bred and grown in large scale propagation nurseries and some have significant travel miles associated with them, especially those that are not particularly climate suitable – and these are not hard to find at major nurseries. Smaller nurseries are more likely to be able to give you sensible advice about the plant you are buying (if you ask!), but are also less likely to be stocking completely unsuitable plants. Many smaller nurseries have their own growing area out the back. Being small businesses, they rely on repeat customers, so they want their plants to do well in your garden so you will come back and buy more.

If you buy plants through mail order, be aware that they can have high travel miles and may not suit your climate. I'm personally not a fan of mail order, although I know many gardeners are. For buying things like flowering bulbs, it is definitely the most economical way to do it. Bulbs are often advertised in March here in the subtopics so that there is time to chill them in the fridge for six weeks before planting them. This is not sustainable gardening, nor to my mind is it very sensible gardening. I knew one lady who went to get her tulip bulbs out of the fridge to plant and found them gone. Her son had cooked them thinking they were garlic. He complained that the garlic had no flavour. Given that tulip bulbs can be poisonous, he was lucky that flavour was the only shortcoming of that meal!

Support your local nursery but remember there are other ways to source plants sustainably. Sharing cuttings and spare plants with friends and neighbours saves on travel miles, ensures the plant is locally suitable and from local genetic stock, and saves the intense water and fertiliser use in commercial plant growing, not to mention the reused pot it has been put into. Sharing plants is also a great joy. Visit any older gardener and they will regale you with stories of who gave them the cutting that is now that glorious shrub you have been admiring. Gardens become places filled with loving memories, as well as plants. Just today a friend showed me a picture of eucharist lilies flowering in her garden and told me that they came from her grandmother's garden, and that flowers from the original clump featured in her parents' wedding. A garden of shared plants can almost become a scrapbook of memories in 3D.

This is not all bad news for the nursery industry. The more plants we are able to propagate ourselves to fill our garden, the more likely we are to go and buy a few more special things to go with our homegrown garden. I have seen people turn away from gardening due to the cost. They were trying to fill a garden with only the plants they purchased at the nursery. Once they realised that they could also purchase plants at their local weekend market or from roadside stalls, or even better to propagate them for free, the released pressure on the hip pocket allowed them to enjoy gardening once more.

Sustainability and budgeting often goes hand in hand. Reduced consumerism saves money and the planet but so does thoughtful consumerism. Buying local, be that plants or anything else, is often cheaper and has less transport miles, in addition to supporting the local community.

Another great way of getting new plants for free is to save the seed of your annuals. Saving your own seed allows you not just a new generation for free, but also means you will still have some of your favourites when they are out of favour with the general public and no longer available in nurseries. There is an added advantage that every new generation grown from seed collected from your own garden is better adapted to your garden, and will be more likely to thrive there. This is especially true of things like peas and sweet peas.

If you purchase seed from cooler climates, it may do ok the first year. If you save the seeds year after year, as the climate shifts gradually, they will be so accustomed to where they are that they will be better able to adapt to keep growing in that climate. Ten years later you may find the seed you have saved each year is still growing beautifully, but the seeds you purchased from elsewhere are struggling.

From a budgeting point of view, I often tell beginner gardeners to stay away from expensive plants. Not only do you have more to lose with a larger investment, but there is usually a reason why some plants are a lot more expensive than others. It could be that they are rarer and hard to find, often they are harder to propagate and just as often it is because they have been transported as large plants long distance to a climate that does not suit them. All of these factors point to a useful rule of thumb that the more expensive the plant, the more chance there is that it will need extra care.

Growing Food and Medicine

We can significantly reduce our carbon footprint and improve our health by growing some food in our garden. It is also a way to reduce our nitrogen footprint (nitrous oxide being a significant greenhouse gas contributing to climate change), as agriculture is the leading source of nitrous oxide emissions.

Many of us have tried to grow food plants at some stage with mixed success. Tomatoes are often the first plant people try to grow, but fail miserably if they plant them at the wrong time of year for their climate zone, or they will plant carrots into soil that doesn't suit them. Other times great rewards are had in the first year of growing food, but the following year it just does not work.

These failures tend to put people off growing vegetables, but it does not have to be so hard. There are two golden rules to growing food at home that especially apply to beginners:
1 – Start with the things that are easy to grow and work your way up to the harder stuff.
2 – The more you take out, the more you must put in, so keep up the soil care work (vegetables like rich soil, so always add compost between crops).

Easy food

Start with growing what is easy to grow. This is usually leafy greens, regardless of your climate. Obviously, it is not worth growing something that you don't want to eat. Mind you, even if you don't eat it, it is still a plant and that alone is something.

Most of our usual vegetables are rather predictable: carrots, tomatoes, broccoli, potatoes. You need to get the right time of year, and this changes dramatically based on where you live. Tomatoes are for summer in southern states and for winter in northern states where winter is not really cold and summer is about fruit fly. You need to watch out for pests. Summer down south is all about cabbage white butterflies on the broccoli, while up north these pests exist all year round. Have you got a plan to save your precious crop? Carrots need fine deep soil that few of us have naturally. Have you figured out yet that growing vegetables is not quite as easy as they make it look on TV? But it is also highly rewarding when you get it right. Getting it right is much easier if you start simple and make a point of learning as you go.

Some of the easiest things to grow are the leafy greens. Getting a plant to produce leaves is a bit easier than getting it to produce tubers or fat roots, or huge fruits. This does not mean trying to get large hearts of lettuce. It means trying the 'cut and come again' lettuce varieties. Try some Asian greens. They are often fast growing and very rewarding. If you live in a warm climate investigate some of the fabulous perennial greens such as sambung, Brazilian spinach or aibika.

Herbs are often also a great easy place to start. Rosemary is a staple these days, as are basil, parsley, mint and chives. These are all things that can be tucked into a spot in almost any garden and do not need a special intensive care vegetable garden. Basil and mint are likely to do better in warmer weather so prefer summer. Coriander, dill and rocket find summer a bit hot and do better in the cooler months – unless you live somewhere that gets cold, in which case give them a go in the warmer months. If you live in a hot humid climate, forget growing coriander in summer. Instead try the saw tooth coriander which is perennial and from Asia, or papalo, a summer annual from Mexico. You may need to find specialist local growers to find some of the best edibles for your climate, but as more of us ask for these sorts of plants, they become more available.

Another very good reason for starting with leafy greens and herbs is the way we use them. The fresher the better. They do not keep very well on supermarket shelves so are often treated with chemicals (usually chlorine) to keep them fresh longer. Nor do they last very well at the bottom of the fridge. You are far more likely to use herbs in your meals if you have them in the garden and can pick them as needed. The same goes for salad greens.

I personally do not have much luck with anything that needs a lot of care. My garden is a bit on the wild side and things have to fend for themselves. It is the antithesis of a successful vegetable patch, and yet I eat out of my garden every day. I have herbs and edible greens tucked into every spot that suits them. As I am preparing the evening meal, I take a basket into the garden to pick what I need.

For those willing to put more effort into their food growing, you will work your way up to the needier plants in no time. Tending a little patch that is close to the door or path so you can see it every day is not hard. Radishes, cherry tomatoes and zucchinis can be a great first-time crop, and with every success, you work up to something more adventurous. Just remember to improve the soil after every crop and grow things at the right time of year. Australia is a big place with a huge variety of climate zones. When seeking seasonal advice for your local climate please check where the information is coming from. A reference from another part of the country is unlikely to fully grasp your local conditions. If you are in a capital city you probably have a handful of local gardening identities who can help. Otherwise, local radio talkback often includes gardening segments early in the morning, and there are local garden clubs almost everywhere. Don't be shy to ask them for advice, they are usually a collective of people who have been growing in your local area for a lifetime.

Think this is a bit over the top? A supplier from Victoria decided to host an event featuring tomatoes in Queensland a few years ago. They got here and found there were limited tomatoes locally available at that time of year. Trying to grow vegetables at the wrong time of year or in the wrong climate is actually a leading reason why vegetable patches fail.

Of course, some advice can be adapted easily to any climate. My grandfather always said that if it was too cold to put your bare bum on the soil, it's too cold for beans. Maybe if we all sat our bare bums on the soil more often, we would be more in tune with what is going on in our gardens!

If this all still sounds like too much effort, get to know the weeds in your garden. It is highly likely that a number of them are edible. The huge advantage to edible weeds is that they require absolutely no care at all. That is my kind of vege patch! In addition, many of the edible weeds are exceptionally nutrient dense, so not only are they free, easy and fresh, they are exceptionally good for us. Many of the common weeds we have in our gardens are actually there because they have travelled the world with people as a staple food. Whilst they may not be grown commercially, things like chickweed, plantain, dock, dandelion, fat hen, green amaranth, nettles, pigweed and even cobbler's pegs have been staple greens for people for thousands of years. Even if you never grow carrots, there is no reason you can't eat fresh food from your own garden. We do not need to grow everything we want to eat to make a difference.

Another aspect of growing food that is popular is growing fruit trees. Once a lemon tree was a common backyard feature. With the shrinking of modern gardens there is less room for lemon trees. If we are determined, we may be able to squeeze in some fruit trees. Even just one or two can be useful. They like to be well fed and they tend to like lots of sun.

Given that fruit trees do take up precious space, think about what you would use before deciding on a tree. Talk to your neighbours. Would they also like to grow a fruit tree and if you each grow something different you can share the produce. A bunch of bananas will all ripen at the same time, making it hard for one household to eat them all. We often only need a couple of lemons or limes at a time. Mulberries can crop far too heavily for one family to eat them all. All of this can be shared. I have two neighbours with mulberry trees who are happy to share the crop. I have a neighbour who planted a lime tree on the verge for all to share. I've encouraged my neighbours to help themselves to figs, passionfruit, paw paws and other things that hang over my fence. I'm about to harvest more turmeric than I can use, so it too will be shared with neighbours. I had a neighbour with bananas to share, until she moved and the new owners chopped them down. That is the downside to sharing, but the upside is far greater.

When it comes to the backyard fruit tree, oftentimes it stays small and unproductive due to lack of care. Fruit trees like water and they like to be well fed. Site them somewhere they get enough sun and are easy to keep the water up to. All the better of you can have the washing machine water diverted to go straight to them. This has made my custard apple grow in leaps and bounds and has saved my lemonade from otherwise certain death. Refer back to the section on greywater for more advice about how this could work for you.

I have planted an orange tree next to my chook pen, with a mulberry and acerola cherry inside the chook pen. They do not always get enough water here but they are well fed. This did not suit the dragonfruit however. It was too well fed and grew too well with less energy put into producing fruit. The bonus here is the mulberry provides summer shade for the chooks and gets cut back hard in winter, letting the light in. Mulberries fruit on new growth, so give your old tree an annual prune for good fruit production.

Another great way of keeping the feed up to fruit trees, especially heavy feeders like citrus, is to place a compost bin next to them. This could be the black dalek shaped ones with an open base, or even one made from a large bucket with a lid. So long as the base is in contact with the soil, and it has a lid, it will work. You can even dig the base of the compost into the ground a little to make it lower and less visible, or to ensure rats can't get into it. This compost bin is not designed to ever get emptied. The fruit tree will empty it for you, as it grows roots directly into the compost and feeds itself. If you do this, it is a good idea to have more than one compost bin, so that as this bin gets full, you can add your scraps to another bin and let it break down. Grow some nice herbs or flowers around the compost bin to help disguise it and as companion plants for the fruit tree. Fruit trees can be a magnet for insect pests. The best way to combat this is to have plenty of flowers nearby to keep a good diversity of predator insects in the garden ready to go to work protecting your fruit tree for you.

Soil contamination

There is something to consider before growing your own food and that is the potential contamination of your soil. Soil contamination is actually something to consider in urban areas, even if you know your plot is not in a location that was once an industrial or landfill site, although there are plenty of these that are now housing estates.

The Macquarie University in Sydney runs the 'VegeSafe' program which allows Australian households to send a soil sample for testing for lead contamination, free of charge. If you are planning to grow edibles, I do encourage you to participate in this. I was happy to get my results back and see that I do not have soil contamination, especially as there are some rather high levels in my neighbourhood. The focus of this work is lead contamination in urban areas (although all Australians are welcome to send soil samples). The key sources of lead contamination are historic from lead petrol and lead paint. Sadly, almost a third of samples tested in Sydney returned levels too high to safely grow your own edibles in your backyard. In Brisbane and Melbourne it was closer to a fifth of samples tested were above safe levels. If this is you, how are you going to grow edibles? In this case you should look at growing edibles in containers or raised gardens with imported soil. Container gardening and raised beds are discussed in detail in an upcoming chapter.

Contaminated soil does not mean you stop gardening. It means you do not eat things grown in that soil. The best thing you can do in soil with any sort of contamination is to add life to the soil. Microbes are a critical part of soil health and healing. Keep composting and keep growing. Plants will absorb the toxins from the soil and store them in their tissue. This of course does remove the toxin form the soil but it means the plant material is now contaminated. Composting this contaminated plant material on site will keep the loop closed and the contamination will stay put, which is fine if you are not growing edibles and the levels are not so high as to avoid all contact. If you are removing contaminated plant material, put it in the black bin, not the green bin. It will be better in landfill than in a commercial compost system.

Using plants to remove contamination in soil is known as phytoremediation. On a large scale the harvested plant material is burnt to reduce the volume of the contaminated material, making disposal easier. When it comes to heavy metals, it has even become possible to extract those metals from the burnt plant material. A project in Brazil is finding that farmers growing maize and canola on a contaminated mine site can harvest a kilogram of gold per hectare.

If you have lead issues, you can try mustard greens as a means of phytoremediation. All brassicas are good at absorbing lead and other metals, so be aware of this if you suspect you may have contaminated soils.

The medicinal garden

Before the days of modern medicine and big pharma, plants were the primary source of health and healing. Growing your own medicine was part and parcel of growing your own food. With the witch-hunts came the ostracization of plant cures and herbal medicine in favour of the men of God and science. This was a time when you went to a barber for a haircut or for surgery (hence the red and white striped poles) and when knowledge of healing plants could see you brutally executed. Happily, times have changed!

An enormous amount of modern medicine is still based on plants. It is estimated that approximately 40% of our modern pharmaceuticals are derived from plants. These days that means isolating chemical components from a plant and then producing it en masse, often in a synthetic form. However, within this process is the acceptance of the healing powers of plants,

Earth Repair Gardening

and much herbal lore is now being scientifically proven. This sort of scientific research is exceptionally expensive therefore less study is put into whole plant use than is put into extracts that can be commercialised to pay for the process.

Fortunately for us, when it comes to using healing plants, we have hundreds even thousands of years of history of use to guide us in safely and successfully using plants as medicine. Our gardens can become our medicine cabinets once again.

Lavender is great for headaches, insect bites and stress relief. Rosemary is wonderful for mental clarity and in a bath for sore muscles. Parsley freshens the breath. Aloe vera heals wounds and soothes sunburn, comfrey heals broken bones and sprains. Geraniums help stop bleeding of cuts. Lemon balm soothes the gut and is anti-viral. The list is endless.

Importantly, there are so many plants with so many simple and safe uses that there will be something in every garden that will be of benefit, if only we know where to look. Many herbalists believe that the plant cure that is needed can be found nearby. Certainly, in nature an antidote plant always grows near a stinging plant. If you are stung by nettle, look around for chickweed, dock or fern fiddleheads.

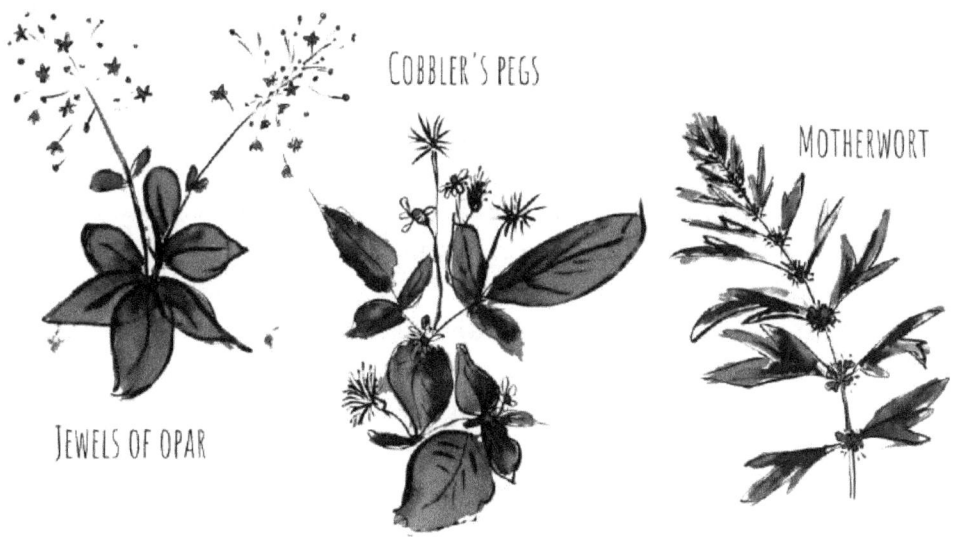

The other important aspect of this is that there will be local medicinal plants, be they natives, weeds or just our favourite flower. We do not need to desperately strive to grow something which does not suit our local conditions, because there will be a plant that does that job equally and will grow for us. This sort of localised information can be hard to find. Your local herb society (if you have one) will be beneficial, as will local Indigenous rangers or local weed experts.

Before using any plant medicinally, do some research. Make sure you know what the plant is and how best to use it. Not all plants are safe to eat. Even some of our hugely successful modern medicines come from highly toxic plants, so always treat plant medicine with respect.

That being said, a pot of aloe vera growing in a hot sunny spot is a must have for every garden I believe. This plant is gold for skin care and especially for sunburn. Given the incidence of melanoma in Australia we really need to take better care of our skin. This includes not only reduced sun exposure, but safe and effective natural skin repair and care without chemicals. This is easy to do with aloe vera straight from the garden.

There are so many plants that are incredibly safe to use medicinally. Many of our culinary herbs are also highly medicinal. We know them as edible plants so we know they are safe to use. A good herb book will be very beneficial if you wish to make the most of your garden medicinally.

Some of the ways I use my garden medicinally are:

- Sida retusa (*Sida rhombifolia*): chew a few leaves each day for sinus, runny nose or unhappy tummies. It can be used as a one off when we are feeling blah or regularly for chronic sinus or tummy conditions.
- Comfrey (*Symphytum officinale*): blended with aloe into a green mush for sprains and broken bones, even aches. It can be spread straight onto the skin and held in place with a bandage.
- Cobbler's pegs (*Bidens pilosa*), plantain (*Plantago lanceolata*) and sida retusa: put a generous handful of leaves into a tea pot and add boiling water to make a herbal tea for colds.
- Dandelions (*Taraxacum officinale*) and nettles (*Urtica sp*): make a great tonic, taken as a herbal tea or just added to meals by the fistful.
- Chickweed (*Stellaria media*): is good for colds and general good health. This can be made into a herbal tea but it is so easy to eat that it can be used in salads, on sandwiches and added to meals. It can even be added to soups, as can so many of these fabulous greens.
- Motherwort (*Leonurus cardiaca*): for stress and regulating female hormones. This is a very bitter and awful tasting herb, so a couple of leaves before dinner will improve digestion as well as having those other benefits. It is easier to swallow when made into a herbal tea with a bit of honey.

- Nasturtiums (*Tropaeolum majus*): for colds and infections. Just eat three or four leaves a day when you are under the weather. They have antibacterial properties, without harming gut flora.
- Basket plant (*Callisia fragrans*): for headaches and muscle aches. This plant is good for so much more as well but just eat a young juicy leaf when you have a headache. I find it works well if I am rushing around with a hangover.
- Mallows (*Malva neglecta*), chickweed and plantain: for skin creams. I have a recipe for the most fantastic healing skin cream in my book, 'Working with Weeds'.
- Geranium (*Pelargonium sp*) leaves: make garden bandaids. If you cut yourself, crush a geranium leaf and wrap it around the cut. It will stop the bleeding quite quickly and start the cut healing.
- Sheppard's purse (*Capsella bursa-pastoris*): for bleeding noses. This little weed is actually good for all sorts of bleeding issues, just chew a few leaves as needed.
- Garlic vine (*Mansoa alliacea*): in the bath for aching muscles. It smells like garlic but a few drops of lavender oil disguise this. Even if it is not overly effective, laying in a hot bath with purple flowers and lavender is divine and relaxing for both the mind and the muscles. You can also add some fresh rosemary or lavender for muscle relaxation.
- Jewels of Opar (*Talinum paniculatum*): for vitality. This weed is both pretty and prolific, so can be added to dinner in great handfuls. No need to do anything except enjoy eating this to get great benefits.
- Aloe vera (*Aloe barbadensis*): after sun or wind, or for any skin complaints. Just pick a leaf, split it open and rub the gel on. You can also use it in your hair as a dry shampoo.
- Chrysanthemums (*Chrysanthemum morifolium*): for stress. Make a cup of tea with the flowers and sit in the garden alone to drink it.
- Mother of All Herb (*Plectanthus amboinicus*): for calming and to improve sleep. This fabulous herb is so easy to grow, and so useful. I use it in place of the Italian herbs when cooking. I use it in place of chamomile for sleep, I chew a leaf to relieve sore throats, and I rub it on my skin to stop mosquitos biting.

I try to make a point of having 10 different leaves in my salads or meals as often as possible to give diverse fibre for good gut health. I can achieve all of this from my garden without putting in any effort at all, but just using what is there.

Native Plants

We are often being told to grow more native plants to be more sustainable. The reasoning for this is usually based around a combination of conservation and reduced needs for watering and feeding. I don't buy it. I have managed to kill as many native plants through lack of care in my garden as I have exotics. A huge number of native plants are simply not suited to garden situations. On the flip side, a huge number are suited to garden situations and they are a joy to have in your garden.

Again, this is an area where, like growing vegetables, a little information can lead to failure. A little more information and you are on the road to success. There is a common misconception that natives are the best plants for a low maintenance garden. Surely since they grow here in the wild without help, they will grow in our gardens without help? Not necessarily.

Firstly, a native plant is one indigenous to Australia. Australia is a big and hugely diverse place! A plant from Western Australia with light sandy soils and low humidity is unlikely to be happy in Brisbane with heavy clay soils and high humidity. Whilst we can overcome this by planting species that are native to our local area, this can also have its challenges. In fact, this can be even more difficult than simply sticking with the popular plants and cultivars available in most nurseries.

Many of these local species are delicate plants with very specific requirements. These are not plants for the inexperienced gardener. Many of these are even challenging for the experienced gardener! A garden situation is a highly altered landscape, it is not the same soil and climate conditions that existed prior to clearing and development of the land. Plants that you see commonly in your local wild areas are more likely to work in your garden than the very interesting but much rarer ones we may be tempted by. When seeking local native plants try and find a local bush care nursery that specialises in growing only plants native to the local area. An advantage of the local bush care nursery is that the staff usually have good knowledge of their plants. If they advise you that the plant is reasonably common and easy to grow, then I advise you to grow it in your garden. If you want to try something more specialist, then talk to the staff about the conditions that plant will require. Specialist native plants are just as possible and rewarding to grow as anything else, but they are not low maintenance.

Many native bushland plants are not used to growing in rich garden soil. Rainforest plants will be better adapted to garden conditions. If you wish to grow a lot of native plants, go easy on the soil improvement. Australia has naturally poor soils in many of the areas that our iconic garden natives come from. These plants (think grevilleas, banksias, hoveas, leptospermums, bottlebrushes and wattles) will not enjoy rich soils. Improve the drainage of these soils, but don't add too much compost. This might sound like just the ticket for those of you who don't wish to be too fussed on the work of soil improvement.

Even without soil improvement, the soil in your garden is not the same as local native conditions. Chances are your soil has been highly impacted by land clearing, leveling, and then turned upside down by building your home. The delicate balance of what was there has long been lost. Chances are you will need to do some soil improvement to make it worth growing anything at all in it.

While natives might sound like a good 'plonk 'em in and forget 'em' option, they are not if you want a nice garden. The thing to think about here is, are you growing a garden or recreating a patch of bushland? If the later, sure, put them in and leave them to their own devices. If you are wanting a garden setting, we tend to want plants to be neater, bushier and flower more heavily than they will naturally in the bush. That means some soil improvement, pruning, and caring for them. This is not high maintenance gardening, but neither is it 'no-maintenance' gardening.

So why grow natives at all? You mean apart from the fact that so many of them are stunningly beautiful and look amazing in any garden? The key reason to have natives in a sustainable garden is not for your sake at all, but for the wild creatures that visit your garden. We tend to think of our gardens as stopping at the fence line, being finite and limited. Our legal ownership might stop at the fence line, but our responsibility to our local environment does not. Nor does nature recognise survey pegs.

When we think of wildlife in the garden we usually think of birds. We have so many wonderful birds in Australia, and the world over people love seeing birds in their gardens. A German study found that a 10% increase in the biodiversity of birds in people's daily lives increased their happiness the same amount as a 10% pay rise. That alone is a great reason to pop in a couple of native bird attracting shrubs or trees.

Here in Australia, many people will pop in a grevillea or two and enjoy the rainbow lorikeets. I am one of these people, and I adore the lorikeets in my garden, as I adore the flowers on the grevillea. I am also keen on other birds and for that I need diversity. Grevilleas only will only get the rainbow lorikeets and noisy miners. I've added bottlebrushes for a greater diversity of small birds, plus native grasses for seed eating birds, dense shrubs for small birds to hide in and native trees for larger birds. While much of this can be achieved with the right sort of garden plants (the spinebills are happy with the large salvias and aloe flowers), adding native plants into the mix will be important for maximising the diversity of wildlife you can attract to your garden. The scarlet honeyeaters will only visit my garden when the native bottlebrush is in flower. No bottlebrushes equals no scarlet honeyeaters.

I am also keen to attract a diversity of insects to my garden. The insects may not be as showy as the brightly coloured birds, but they are just as if not more important. I have had over 20 different butterfly and moth species in my garden this summer. There are many more than this, but these are the ones that are showy enough to be noticed and easily identified. I have a variety of different bees and beetles, even sometimes fireflies, which is exciting. There are wasps, lacewings, damselflies, hover flies, bee flies, ladybugs, assassin bugs, katydids, grasshoppers and spiders galore. It is on this mini beast level that we can find the most abundant and fascinating diversity. Some amateur nature watchers are recording over 300 different insect species in suburban backyards here in Australia. None of these creatures know that a garden is not native bushland. They are simply seeking suitable places to feed and breed. As so many of our insects are reasonably generalist in nature, they are easily catered for by creating a garden that is diverse in flowers and leaf types, and perhaps a little untidy so there are places for them to nest. So long as we are growing a good variety of flowers, they really don't care if they are native or not.

Some of our insects however are a bit fussier. The Richmond birdwing butterfly caterpillar can only feed on the *Pararistolochia praevenosa* vine (and are poisoned by the introduced relative, Dutchman's pipe). Christmas beetles, like koalas, prefer to eat eucalypt leaves. We tend not to realise that such specialised relationships exist until they are broken. It is in studying the decline of the showy insects we love that we find they are far more selective than first realised. Chances are there are many more of these specialised relationships than we will ever know about. By planting local natives in our garden, amongst our other favourite plants, we have the opportunity to appreciate them and attract a greater range of local native wildlife. A recent study found far greater risks of extinction amongst Australian butterflies than had previously been thought. Many of the butterflies identified in this study had very specific relationships, not just with the host plant, but many of these butterflies also depend on a relationship with specific ants. While some of these butterflies were at risk due to agricultural clearing, many were in serious decline around capital cities and locations of urban population growth. This suggests we can play a role in species conservation by ensuring we include local native plant species into our gardens as a matter of course.

Even by adding just a few local native species to your garden you can help create linkages with local bushland remnants. These linkages may be small, but the more we do this, the more we create a greater habitat range than exists in small, isolated bushland reserves, and create more genetically secure populations of wildlife, great and small. We know that especially in urban areas any bushland remnants tend to be small and degraded. We know they really ought to be bigger to be viable. We actually can make them bigger by adding just a few native plants to our gardens and making our gardens safe and attractive places for wildlife to visit. Whether our garden is largely exotic flowers and vegetables with just a few natives thrown or an entirely native garden, it is all contributing to the extension of habitat corridors.

At this point I am sure there is more than one avid vegetable grower wishing for less wildlife in the garden. It is not true that all native animals prefer native foods. Anyone who has tried to plant native food plants to entice possums away from their crops knows this. It simply means more food and more possums. Dealing with the local wildlife we don't want is always going to part of gardening, whether you grow native plants or not. In fact, the more pesky and troublesome the local wildlife, the less they are dependent on native food plants and the happier they are to eat anything you may wish to grow. Some animals (and some plants) are highly opportunistic. They are usually generalists, meaning they are not fussy eaters. These are the ones we get to see and appreciate the most as they are adept at co-existing with humans. They might eat a few plants, but usually they are a bonus to have around. Adding more local plants to your garden won't impact much on the presence of these common wildlife, but it will increase your chances of having a greater diversity of other wildlife visit.

We started this chapter by mentioning that natives are not no-maintenance gardening. They will need watering to get them started at least. In dry years all natives will do better with top up watering if it is available. With natives as with any other plants, be aware that we want deep roots, so make sure you water deeply and not too often.

Natives will also need pruning. Most natives will do very well with an annual tidy up. Some, like the westringias, smaller bottlebrushes, eremophilas and leptospermums can even have a light trim a couple of times a year to keep them compact and dense. Those that are woody shrubs need a bit more care when pruning. Plants like grevilleas and banksias can look awful in gardens when pruned badly. Pruned well and they are brilliant garden plants. I'll let you in on a trick for getting it right. Identify the branch to be shortened and look at where the shoots are along the branch. Once you cut the end off the branch the next shoot will become the leader shoot. The direction that this shoot is facing will be the direction the new branch grows. If that shoot is facing down, or towards a path or otherwise in an awkward direction, you are creating future problems. Follow the branch back until you find a shoot growing in a direction that will be good for the shape you would like the shrub to be and cut just above it. This shoot is now the leader shoot and will determine the shape of the new growth.

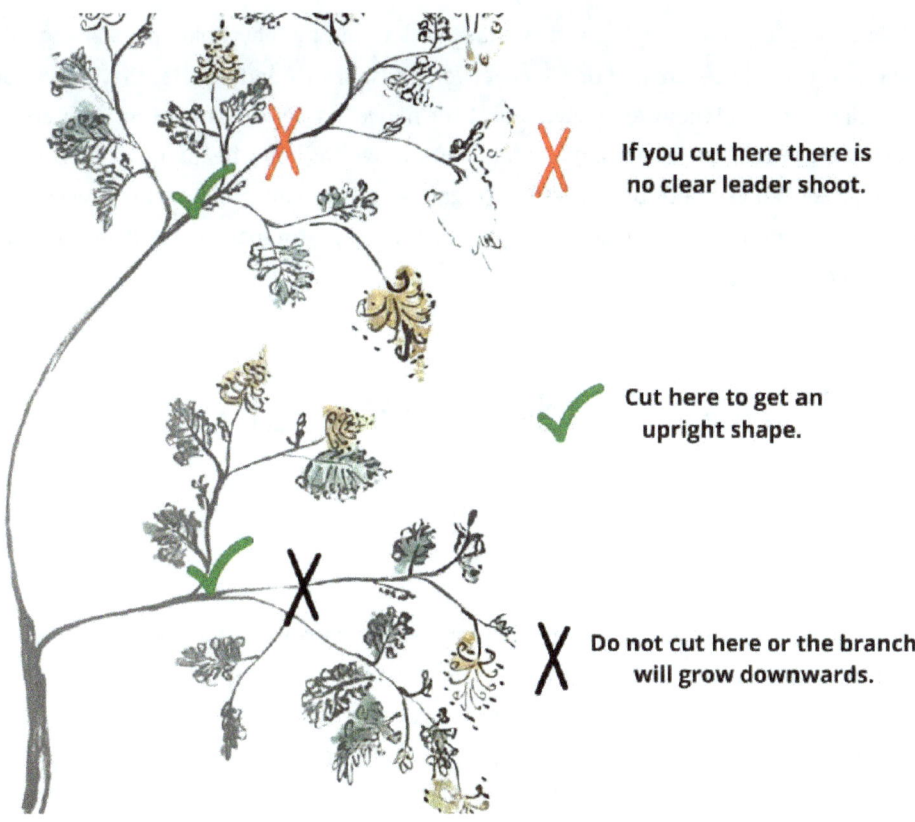

When pruning a grevillea or other woody tree or shrub, it is important to prune in such a way as to allow the plant to grow back into a desired rather than a distorted shape. The first bud behind the cut will be the new growing shoot. The new branch will grow in the direction of this shoot. By pruning back to a bud that is pointing in the direction that you want the plant to grow, your pruning efforts will result in a well-shaped plant. Don't cut back to a downward facing bud unless you want your plant to grow downwards. Generally, we are looking to prune to upward facing buds to create a plant with upward reaching branches.

Yet another group of natives, the strappy plants including lomandras, dianellas and kangaroo paws will all do well with a big chop after flowering. They can be cut in half or more, even to just 10cm tall. This doesn't just tidy up the plant, removing all the old and tatty leaves, but it encourages fresh new growth. Do the same for any of the native grasses you are growing to keep them looking great. It doesn't always need to be done every year, but every couple of years will make a big difference in keeping the plants looking good and growing vigorously. These strappy natives are well worth growing. Lomandras (followed by dianellas) are the absolute best plants for stabilising erosion prone banks. There are varieties for full sun, full shade and all sorts of conditions. There are a range of ornamental cultivars with different foliage forms and colours. They have extremely dense root systems. I love using them to slow runoff, trapping organic matter as they slow down water. They hold the soil together, preventing erosion; they provide dense cover for frogs, lizards and insects; they flower and attract pollinators - and they are tough to boot! The native grasses can also be great for erosion control. Native grasses are important habitat plants, as well as being beautiful accent points in a garden. They can be used to great effect to give a soft romantic look as they sway in the breeze and are important for small seed-eating birds and also for many butterflies. Convinced to add a few to your garden yet?

Tania's Garden – Wild at heart

Tania's garden is the sort of place where you can really escape from the world around you. It is a small suburban garden (609sqm) that is so full you can completely disappear into it. Every space is a canvas for found objects and garden art, all with plants tucked into them. A garden where imagination has been given free reign and hours could be lost to the joy of it. The children over the road love to come and explore here, but so do adults. It's the sort of place where your inner child comes to the fore and does not want to leave!

Tania openly admits that her inner child is very much alive and well in her garden. She grew up in a flat, dreaming of having her own garden and has certainly made up for lost time. The garden is an adventure of items that are fun, curious and many that hold a special meaning. There is a collection of large seashells dotted throughout the garden. These came from her great aunt many years ago, who was given them by her father. Tania's great grandfather had collected the shells on his seafaring travels in the late 1800s so they evoke a sense of history and adventure.

Not all the fun items hold that much weight. A skeleton hand on the shed was broken off the model by medical students where Tania worked, so instead of binning it, she put it in the garden. There's an old toaster that she found and just loved the retro look of, so it became a planter. Some items, like the collection of dragon sculptures made by children, Tania purchased just because she loved them. Other quirky features were created as a way of reducing waste. Sections of old concrete that had to be removed have been repurposed as garden edging. This does not sound very appealing, but with the addition of a few glass beads and time for moss to grow, that same piece of old concrete now is full of character.

While the garden was always intended to become a quirky and arty sort of wonderland, this was not Tania's only goal. The very first thing she planted when she moved in, even before she unpacked, was a wattle. This fast-growing small tree provided shade from the afternoon sun, and paved the way for a garden bursting with wildlife. Native plants have been tucked in amongst exotics to create a perfectly harmonious union. Salvias and banksias work surprisingly well together. Tania finds the natives to be highly seasonal. When each is in flower they attract an abundance of birds and bees and are a joy.

The exotics provide the year-round flowering that keeps the birds and bees there even when the natives are not in flower. The brown honeyeaters like the pink monkey tails shrub, and the cardinal creeper has blue faced honey eaters in it most mornings. The garden is very much alive, with the buzzing of pollinators everywhere, but there is as much hidden as there is seen when it comes to wildlife. Tania has very deliberately avoided neatness in the garden. At ground level there are sticks, mulch, interesting objects and interesting plants, all creating places for lizards and frogs to hide. There are bowls of water at different heights and depths to cater for various critters. A pile of old breeze blocks with a water bowl on top turned out to be the perfect home for tree frogs. A blue tongue lives under the old concrete laundry tubs (planted up of course!). There are lots of lizards in the garden in summer and Tania puts bits and pieces around everywhere to help create nooks and crannies for them to hide in.

As much as Tania loves the wild visitors, she has her own feathered and fluffy friends to also cater for in the garden. She has two indoor cats with access to the area under the house via a cat enclosure. The two old dogs, Peggy and Elsa are beautiful company and love following Tania around the garden. They are not at all interested in chasing lizards or birds. Given that they are in turn kept company by a couple of bantam hens, they are pretty used to feathered company.

With all of these creatures sharing Tania's garden, it is very important to her that it is a safe space for everyone. To that end, chemicals are an absolute no-no in this garden. Compost is the main means of feeding the garden. There seem to be compost bins everywhere, but they are so lost in the overgrowth you barely notice them. Tania didn't set out to collect compost bins, she simply accepted them as they were offered to her. Her plan is to fill a bin in new sections of the garden to improve the soil before planting. Well that was her plan, bare spots in which to create new sections of the garden have long since run out but compost bins are still working to improve the soil in the garden.

Tania collects all sorts of organic matter to feed to compost bins. In addition to her kitchen scraps she adds horse manure from local stables, shredded paper, bedding from the chook pen, the neighbour's grass clippings (so long as they haven't been sprayed) or anything else she can get her hands on. She is particularly keen on adding roadkill to the compost. It is usually birds and possums, but she did bring home a hare once and struggled to fit it in the compost bin. She has learned to ensure the corpses are well covered to keep the smell under control. The deceased animals get a bit of a burial this way, with fresh flowers and herbs added to the top of the compost they are buried in. Dead bodies are a source of organic matter and nitrogen which plants like. Many of us bury our beloved pets and plant a special plant on top that tends to grow well with the rich compost. For Tania, it is a way of honouring the wildlife killed on our roads at the same time as feeding her plants.

In addition to the compost Tania does use a bit of organic fertiliser and rock minerals. She also likes to give the garden a foliar feed regularly which really helps for all the small plants in pots, of which there are many. Most of the pest control work in the garden is done by the chooks, the lizards, frogs, spiders and other predatory insects. Very occasionally Tania needs to intervene with a molasses spray.

While the garden is designed to be chemical free and completely safe, sadly Tania does struggle with neighbours who do not share this ideal. She has a bare strip along one side fence to ensure that neighbour does not get upset about the garden and become tempted to spray along the fence line. Her rear neighbours use a lot of aerosol chemicals in their hobby motorbike restoration activities. This is much harder to block. Dense planting along the back fence of things like tiger grass help to create not only a privacy screen but to trap the worst of the chemical drift.

To keep weeding and maintenance down, pathways through the garden are all created with woodchip mulch from local tree loppers. As this compresses with foot traffic it becomes an easy surface to walk on while remaining porous and allowing water to penetrate. It also reduced the need for hard garden edges and the feeling is that the pathway has been woven into the garden, rather than that the garden has to stay put in its beds. Another advantage of the woodchip pathways is that seeds can drop and grow there. Some of these things, like marigolds, are allowed to stay; others like Egyptian spinach are moved to more suitable spots in the garden.

Every now and then an ornamental plant starts popping up as seedlings in the pathway with too great an abundance to be ignored. There is a wetland reserve at the end of the street, so Tania is very aware of the potential of weeds to invade native bushland. If anything looks like it is getting too weedy, it is removed from the garden for the sake of the integrity of the wetland.

Another aspect of sustainable gardening that Tania has embraced is growing food. She does not grow a lot, but there are fresh herbs for the picking, a very productive passionfruit vine, dwarf bananas, paw paws and a variety of edible leafy greens. Tania actually grows the paw paw for the fruit bats. If there is any fruit left for her it is a bonus. This truly is a garden for the wildlife!

Pots

You cannot be a gardener and not have pots. We love growing plants in pots, we buy plants in pots and we share plants in pots. I have yet to meet a gardener without a pot or two lying around.

When it comes to pots, we can divide this into two broad distinctions – the practical and the decorative. The practical are the ones most likely to end up in the bin, so let's start there.

Plastic pots

We usually buy plants from nurseries in plastic pots. I am reusing as many as I can in my home nursery, taking them back to nurseries that will accept them (most won't), and the rest eventually go in the bin. Most of the time, these are seen as single use disposable items that end up in landfill. When it comes to recycling these plastic pots, there are some success stories, mostly around retail nurseries with collection bins especially for pots. Most municipal recycling systems will only accept the coloured pots. For some reason the black pots (and most of them are black!) do not get detected by the sorting machinery and are rejected to landfill.

Sadly, nurseries with any sort of pot return option are exceptionally few and far between. Victoria seems to be leading the charge here. I have successfully given away boxes of unwanted plastic pots to someone who can use them via online marketplaces. Being able to giveaway your unwanted items via an online local marketplace is fabulous and a great way to not only get rid of junk but also keep things out of landfill. Good as this may be, there are still far more plastic pots out there than anyone can possibly use, so they will remain a landfill issue until we address this more seriously.

There are a few biodegradable pots starting to appear on the market. These are usually a naturalistic dark beige sort of colour. The colour is just a marketing ploy, it is not that they are naturally produced in that colour. It would seem that they are a more sustainable option – if the plant you want to buy comes in one of these pots. They are designed to break down in approximately 12 months. Unless they are made from plant derived materials rather than petroleum-based plastic, this is of no gain. Plastics that break down may seem like a benefit, but in fact they only break down to tiny particles which are even more environmentally harmful. Microplastics are very stable and persistent. They get through most physical filtration and screening systems and end up in our oceans, to become a very persistent part of the food chain.

So, while we may be drowning in a world of disposable plastic, there are currently very few options when it comes to buying plants. Shopping from local growers can be beneficial, as local plant people are more likely to be a small home-based or community run business which will happily take pots back for reuse. I know of a couple of local plant growers who have reused pots many times over. Of course, these businesses offer a few added bonuses – they are local so reduced transport miles and the plants are far more likely to grow in your climate, not to mention you will be supporting a local family.

Having your own mini home nursery is a very sustainable way to keep up plants for your garden and to share or swap with others. It is also a way to reuse some of those plastic pots, even if just on a small scale. Don't be tempted to buy new plastic pots for your home nursery. There will be someone not far from you with a supply ready for the bin if someone like you doesn't take them away for free.

What is a realistic solution to the plastic pot pile up? At the moment there isn't really one available, but hopefully that will change in the next couple of years. A grant has been given to set up a national industry led plant plastic product stewardship scheme. This is intended to include recycling of not only plastic pots but also plant labels, trays and stakes. This project is not due for completion until 2023, and even then time will be required for the roll out and public uptake. It is wonderful to see, but this is just the very beginning of the change that is needed. Change is driven by demand for it, so as the primary consumers of plastic pots, we are the ones who should be driving that change by asking for it.

Ask your local nursery to take back the empty pots, or even just return them anyway and ask them to find the solution. A few years ago activists started returning plastic wrapping to supermarkets and leaving it there with the claim that they sold products in this and therefore they were responsible for it. This created a problem that had to be solved and now most supermarkets have a soft plastics recycling collection point. Increased pressure from us is going to see additional drive to get this project happening and successfully taken up by as many nurseries as possible.

There are currently some recycling facilities in Victoria that do take back used pots and recycle them – into new plastic garden pots. Since when did this qualify as recycling? Recycling is about using the waste material and turning it into a new product. I suppose they are doing this, only it is a new product that is exactly the same as it was before. Surely we could do better by starting with a washing and sorting facility? I suppose this is too labour intensive and is far easier to chop all the plastic pots up together, melt them down into little plastic pellets at high temperatures, then injection mould at high temperatures into new plastic pots. It is energy intensive but energy is cheaper than labour. It is better than sending the pots to landfill and making new ones so perhaps that is enough.

It would be great to see further development of non-plastic alternatives such as bamboo or cellulose. There are companies around the world making plant pots from natural materials but it is a fledgling industry so currently they are an expensive option with limited choice and high travel miles. Bamboo waste from furniture making is being moulded into plant pots which look very much like our plastic versions. They are as durable, which is great for the nursery industry, take around three years to break down and can be composted.

Cellulose pots are made from paper and cardboard pulp. Their useful life is around three months, which actually makes them a perfect substitute for seedling punnets. A bonus here is that they can be planted directly into the ground to avoid damage to roots, as so often happens when we try to get seedlings out of plastic punnets. The downside to these is that they are often bound with a form of plastic, so not as fully biodegradable as it would seem. Still, if we keep asking and create demand, we will see change towards a product we are happy with.

There are some options already available and becoming popular. These include the 'Jiffy' pots, which are made from compressed peat moss and wood pulp. They are only available in small sizes as they are designed to raise seeds and cuttings, and to be planted 'pot and all' into a larger pot or the ground. There are also similar pots around made of coir (coconut fibre). Unfortunately these tend to dry out too quickly and their shape easily falls apart so they are not yet a viable option. Do avoid the planting 'pellets' at all costs. These may seem like a nice easy way to grow seeds, but are very troublesome in the long term. The pellet is held together with a plastic mesh, which can take years to break down. The mesh is very restrictive to root growth. Plants tend to do well in these pellets until the roots fill the space. When the roots cannot expand out of the pellet the plant languishes and dies. They are used extensively in the plant propagation industry. If you find your new plant languishing check the roots – if it is planted in one of these pellets, re-pot it, tearing aware the plastic netting to set the roots free.

We are currently seeing the phasing out of single use plastic in the food industry, with bamboo and cardboard being very successful alternatives. The more this industry develops, the more opportunities there will be for it to flow over into the horticulture industry. As consumers we need to do our part and keep asking for it.

Decorative pots

The market for decorative pots is pretty huge. They can be expensive to buy, and yet they are still so often treated as a fashion item rather than a forever item. I personally have never bought a decorative pot (unless I wanted the plant in it on a discount table). I simply have no need to. Other people throw away so many great pots on kerbside rubbish collection days that I have an endless supply. It is amazing to me how much of my garden - plants, pots, ornaments, furniture, landscaping materials all came from something someone else was throwing away. I am a HUGE fan of kerbside rubbish collections. We live in a society that encourages consumerism, which in turn leads to a lot of great stuff, with plenty of life left in it, being updated and replaced well before they need to be. It is wasteful, expensive and unsustainable, but it is also a bonanza for those of us who love to collect and repurpose what is no longer wanted by someone else.

Something as simple as a nice pot with a chip in it can easily be dealt with. Turn the chip away from the main view or grow a plant to cascade over it. A style of pot that is out of fashion, such as an old terracotta pot, might be a little harder to disguise. Simple terracotta pots never really go out of fashion, and over time get a glorious, aged patina that money can't buy, even if that is not to everyone's taste. Lichen, stains and dirt can all be scrubbed off to make the pot look fresh again if you prefer. Some of the patterned terracotta pots do have quite a dated look about them. Use these ones in a cluster of pots so they are partially hidden by the group. This makes the pattern less obvious but uses the pot to enhance the look of the group.

Old concrete pots with their hideous patterns are becoming highly sought after. Due to their weight, they are less often thrown away. It's easier to smash them and move the bits than to reuse the pot by moving it elsewhere. Many of these pots date back to the 1930s – 50s so are well and truly vintage and increasingly collectable. If that style is not to your taste, you will have no trouble finding someone willing to take it off your hands.

Any of these old pots can be given a new lease on life, and a new look, simply by giving them a lick of paint. I've even used some leftover bits of chalk paint to give a bit of character to old plastic pots and make them look better in a plant stand. For those who are keen to repurpose, you will be amazed at what sort of things make great pots. The inner drum of washing machines for example, or the barrel of old cement mixers, wheelbarrows, old metal buckets, old metal chests, coppers, kettles, tea pots, saucepans, even china cups and old tins. There is a bit of a knack to arranging these items in the garden so that it does not end up looking like a garden in a rubbish heap – try and create a theme so that the items all relate to each other in some way. Vintage items can be easier to style with, as they have a charm all of their own.

Vintage items can also be harder to get hold of for free, so you may find yourself paying for this look. Olive oil tins can work well and look great, if they are perhaps pared with baskets or timber crates. They can look like a pile of rubbish if pared with polystyrene boxes or plastic buckets. Think about drainage. If you can't drill holes into the item, place it under cover and water it sparingly. Or grow water-loving plants in it.

I am a huge fan of repurposing items in the garden and love finding creative ways of planting into different items (stay tuned...this is the topic of another book!). I do draw the line at old toilets. The shape makes them work well as a pot, but there is just no disguising the fact that it is a toilet, and they are objects without charm! When I have been designing gardens for others, I have had to purchase new feature pots. In these situations, think through the purpose of the pot and its role in the garden so that you can make a choice for the best possible pot to fit the purpose.

As with all landscaping materials, choose pots to fit to the design, longevity (of both style and lifespan) as well as providence. Is it locally made? Does it include recycled materials? A terracotta or clay pot will dry out quickly – will this be problematic? Unless you are growing very drought hardy plants it could well be. A glazed pot may be better for this purpose. Would a fibreglass or recycled plastic pot be more practical and still beautiful? If the potted feature is for a balcony then weight may be a consideration, in which case the fibreglass or plastic (hopefully recycled) pot will be better choices.

Is the pot you have chosen big enough for what you wish to do with it? This relates both to the plant to go in the pot, and how the pot will fit into the space it is intended for. If you are creating a long-term potted feature for the garden, go for the largest pot size you can. A pot that is too small for the space tends not to look right and will probably be thrown away for a design change.

A pot that is too small is also a common problem for the plants in the pot. While a smaller pot effectively acts to bonsai a tree and keep it as a small pot plant, if it is too small, there will be too little root space and the tree will be too small. This is especially problematic of growing fruit trees in pots. A larger pot allows you to plant multiple plants in one pot, creating a potted garden, not just a potted plant.

Lots of small pots are more work to care for than a few larger pots. The smaller the pot size the faster it is likely to dry out and need more watering. It will also use up its supply of nutrients in the potting mix faster and need more feeding. All of this needs to be factored in to creating a feature that works, instead of finding yourself with a feature that needs replacing. By thinking these points through first you are more likely to have a happier long-term relationship with your pots.

Repurposing in the Garden

Repurposing is a fairly new term, and is part of the pyramid of 'R's that we are all becoming more aware of as we try to live more environmentally conscious lives. Repurposing simply means using an item for something other than what it was made for. This can include alteration of the item such as putting holes into a teapot for drainage to use it as a plant pot, or chopping up a wooden pallet and rebuilding it as a garden table. Repurposing and gardening have long been companions. Cleaning up empty aluminium cans and tying a hessian ribbon around them to reuse as ornamental pots might be very eco-chic right now, but it has been done out of necessity for as long as there have been aluminium cans. The enormous supply of cheap plastic pots that gardeners have access to these days is a fairly recent phenomena, and one way in which gardening is simply keeping pace with our modern disposable lifestyles.

Using repurposed items in the garden is not just about being more sustainable. It is also a great way to add character to the items in your garden, save money and can be a lot of fun. Many older items are full of the sort of character we now associate with a time gone by, even if they are not entirely antique. They are things that may have nostalgic value, or simply a shape or style that is no longer made. Older items were made to last, unlike new items. A metal watering can is a case in point. You can buy lovely looking metal watering cans new at almost any hardware store, but in a year or two they will have rusted through and lose their charm. When you find an old metal watering can, you will immediately know the difference. It may well be 50 years old and have no rust. It will be heavier than you expect and probably won't have a pretty flower stamped on the side, but it was built to last.

It constantly amazes me the things people discard. Most major cities have some form of annual kerbside collection of rubbish that is too large to fit in the rubbish bin. This is a treasure trove of great items to reuse in the garden. So much that is thrown away simply needs a little imagination to see what it could become.

You will probably find items in your own rubbish. Got a new set of coffee mugs? What do you do with the old ones? With some holes drilled in the bottom they may be perfect little planters for succulents or African violets. Time to update the washing machine? If your old model is beyond refurbishing and reselling, take the central drum out to use as an interesting planter. Got given a nice tin with biscuits in for Christmas? The tin can have a hole punched in and be used for planting shallow rooted plants like succulents.

If you have no success in your own or other peoples unwanted items, try the opportunity or secondhand shops. I have already mentioned using all sorts of interesting items as pots, but there are plenty of other ways to repurpose items in the garden.

Reconstructing old pallets

These can look so fabulous. Any quick Internet search will give so many great ideas – vertical gardens, potting benches, tables and stools, even cubby houses. Realistically, the majority of pallets out there available for free are cheap pine wood with a short lifespan. Before you put a huge effort in to your project, check that the quality of the wood is up to the task, or you may find your fabulous new project doesn't last long. Quality hardwood pallets are out there, just a little harder to find, and you may have to pay for them. Of course, how you finish your project will contribute to its lifespan. Painted or oiled finishes are easy to do at home and will extend the life of pallets. Staining your timber pallet creation with old engine oil will also extend its life and give an interesting finish, but it will also make it highly toxic and unsuitable for growing edibles. It is a good thing that very few people can change the oil in their cars anymore. It means that there is much less old engine oil left in garages that can be put to other uses, especially in the garden. There was a time it was saved especially for painting the timber fence. Anything this toxic should definitely NOT get repurposed in the garden.

One really great way to reuse flimsy pine pallets is to stabilise steep sloping gardens. Lay the pallets on the ground and peg them in with small stakes to stop them sliding down the slope. Fill the spaces in the pallet with soil and then plant into these spaces. Over time these untreated pallets will rot away, adding carbon to the soil. As this happens, your little plants have grown enough to be able to hold the soil and stabilise the slope with their roots. By the time the pallet rots away, it is no longer needed.

There are a few other factors that will affect the life of your repurposed pallet, and this is your climate. Hot wet climates are wonderful for creating compost because everything rots faster in this climate, even timber. If you are near the sea, salt spray can increase the speed that things deteriorate. Even placing your item in a shady damp spot in the garden as opposed to siting on a sunny patio can make a big difference. If your item is in contact with the soil it will rot faster. This needs to be factored into your expectations. None of it matters if you are having fun with free pallets and will be happy to create new things in a few years. If this is your masterpiece, you might want to plan for a longer lifespan.

Old china

Old china can be very cute! A simple teapot collection, or a collection of blue and white teacups can make a lovely set of planters for small plants. Use a fine drill bit designed for glass and tiles to carefully put a drainage hole in them. This will be best done with the item placed on a piece of wood to absorb any vibrations that could shatter the china. Drilling through china takes patience, if you push too hard or rush you are likely to break the piece. If you are not able drill drainage holes, place them under cover and water them carefully to avoid wet feet. If you have a few accidents while trying to drill holes in your old china, the broken pieces may well be ideal for a garden mosaic. I have also seen old teacups used very effectively to make cute bird feeders, and teapots hung up in trees as little bird houses. A string of white or clear glass beads from the spout can look like it is pouring water for a quirky touch.

Lovely old plates are good to smash up for making mosaics. There are so many ways that mosaics look great in the garden. We often think of a little table top but I find stepping stones created in an old terracotta plant saucer are a good place to start when setting out with mosaics. If it was a plate you particularly liked, find a terracotta saucer of similar size and reconstruct the plate as a mosaic in the saucer. This can be a fun way to create very personalised stepping stones (especially if it is an entire set of plates) or other garden features for your garden.

I have also created bird feeders from old plates. To do this drill three evenly spaced holes around the perimeter of the plate and hang it with chain. A hole in the centre will be good to prevent water pooling in the plate and rotting the seed (although all bird feeders should be regularly cleaned). The hole in the centre can be used to hang a small ornament or string of beads for added visual interest once the feeder is hung in the garden.

Chairs

Even if a chair is never going to be sat on, just the sight of a chair in a garden makes us think that taking the time to sit is a possibility. Our mind slows and calms at the thought. In this mad busy world we all live in, beautiful chairs in beautiful gardens are moments of pure joy and serenity which are to be encouraged. Chairs can make great pot stands – with or without their seats on. In my garden I have chairs for sitting on, chairs holding pot plants, chairs protecting plants from dogs, chairs as plant supports, and some that swap between these jobs! There are so many old chairs being thrown away that have fabulous shapes. They may be too wonky to sit on again, but not too wonky to look great in the garden. I've put a chipped pot under a broken leg to sit a chair straight and then sat a basket of bromeliads on the broken seat. It looked fantastic. The chair was painted deep pink, and came off a rubbish pile for free. It lasted about five years, by which time it fell to pieces that could be composted and I was ready for a new look anyway.

Insect hotels

Insect hotels are a bit of a craze at the moment and one that I really enjoy. They certainly do no harm and can add a lot of fun and charm to a garden.

You can make them yourself, which in itself is fun. Old CD towers make great insect hotels, and let's face it, not many people are using them to hold CDs anymore. Otherwise think of things like unwanted dolls houses or even old medicine cupboards. A requirement however is to ensure they are solid. Plastic can look tacky but also become brittle and fall apart in sunlight. Chipboard will disintegrate if it gets wet. Insect hotels can be filled with all sorts of fun objects which appeal to you. Leaf cutter bees love pieces of old garden hose cut up to approximately 15cm lengths. I let my son help me make one and it ended up with a few old toys in it, which really only adds to the appeal for us all. Even the insects are happy to find their way in and make the most of the spaces it provides. There is currently a campaign against mass produced insect hotels with the claim they act as deathtraps. This is a bit over dramatic. Your insect hotel will be far more successful if you do a little research into what size holes and how deep they should be to suit your local species. Even if you get all of this wrong for local solitary bees, you will find spiders, ants, skinks and other creatures will put it to use. They are all part of the biodiversity of our gardens and so all welcome too. If a bee decides to check it out and gets eaten by a spider, well it was a death trap for that bee, but this is a risk with any potential nesting spot, artificial or natural. Do however avoid using treated timber as this will be toxic to all the critters who may wish to live there. While you are busy using what you have to create fabulous insect hotels, why not also create homes for other critters? Old PVC plumbing pipe can be put to use in the garden making homes for lizards and frogs. You can be creative and make lizard and frog hotels as well. To make a lizard hotel, place pieces of pipe, logs and rocks into an old bird or small animal cage. Make sure the cage has enough space between the bars for the lizards to come and go. Then place this in the garden in a warm sunny spot. Leave a gap between the top of the logs and other bits and the top of the cage so the lizards can bask in the sun safe from cats and dogs, as well as other predators.

Pieces of PVC pipe can also be used to make frog hotels. Cut the pipe into a variety of lengths. If you have bend sections, these can be included as well, creating a curved top on the pipe. Place the pipes standing upright in a water pot, or any pot without drainage. Fill the pot half way with soil and plant some water plants. Mulch this potted garden with gravel to help keep the water clean and fill the pipes with gravel to the level of the pot. Then fill the pot with water. The frogs will find their own way to your water garden and the pipes make great hiding places for them which remain humid.

Pots

Many found items can make great pots. We have already mentioned things like teapots and metal buckets but let your imagination run wild. Anything that holds a plant and some soil can be turned into a pot. In having fun with things to plant in, there are a few things to keep in mind. Drainage is important – can you make drainage holes in it? If you can't make drainage holes, can it be used for bog plants or a small water feature? How durable is the item? If it is going to rust away quickly it may not be worth the effort. Wooden boxes tend to swell and fall apart when they get wet, although they can look great before they do.

Other items that get repurposed as pots will include things like old Tonka trucks, boots and shoes, colanders, watering cans, and saucepans. I have seen toasters that look fabulous planted up, but I have also seen some terrible ones. I had a great old drum with a broken skin planted up, until it fell over and smashed. Let your imagination run wild! Perhaps the most important question to ask is - does this item make me smile and will it look great and add character to my garden?

Baskets

Baskets are usually easy to find and can look great. They work best with the plant kept in its original pot and then placed in. If you want to plant directly into the basket, line it first with an old t-shirt. You could buy a liner, but while we are repurposing, there is no need to buy anything. Baskets are very free draining, which can work well for things like orchids or hoyas, but they can dry out far too quickly for most plants, hence the recommendation to keep your plant in its plastic pot inside the basket.

Plant choices

Matching a plant to a container which suits it is not just part of the fun, it is a crucial part of the success of repurposing in the garden. Succulents are popular for repurposed containers because of their hardiness, and their shallow root systems. This means they will cope in small containers that dry out easily. Peperomias and spider plants (Chlorophytum) can often also cope with similar tough conditions.

In general, the larger the plant, the larger the container it should be planted into. Depth needs to be considered. Generally old wheelbarrows are very shallow and therefore do not allow for any significant root depth plus they dry out quickly. Factor this into your plant choices and you will have greater success.

Succulents can look fantastic in all sorts of containers but they are not the only choice. Begonias look fantastic in an old wheelbarrow that is positioned in the shade. Flowering annuals can be fairly shallow rooted and can work well in old wheelbarrows as well, but will need to be watered often. You can be as creative with your plant choices as you are with your containers.

Think about the root space available in the container, and what sort of plant will be a good fit – for the root space I mean! There is no point having a plant that looks good in the container on the top if the below ground part is a bad fit. And after all, finding a good-looking plant is the easy part, they are all fabulous. Even something as simple as mondo grass can look great in a fabulous container and set in the right place in the garden.

Theme

Try and keep some sort of theme to your items. This is important to unify the garden. A lot of unrelated items can end up looking like just that – a lot of unrelated items with plants stuck in them. This could work very well if the unifying factor is that they are all planted with the same plant, but we gardeners usually go for variety in the plants before choosing the containers.

A theme could be all kitchen related items, or all old wood and rusty metal, or all machine related, or even just a colour theme. If your items are unrelated but all loved, can you plant them all with a theme in the plants such as different coloured geraniums, or lots of different red flowering plants? If this still does not bring them together as a collection position them so they are not in the same line of sight. The theme may not run through the entire garden, but each view of the garden is not too confusing.

When we talk about theme, this relates to more than just the choice of repurposed containers. This also relates to things like materials used in the garden, the pairing of containers and garden art and even flower or foliage colour. Old rusty gates will look great with things like plants in old metal buckets and watering cans or even old Tonka trucks, but won't be so effective with a collection of gaudy 1970s lamp shades (yes turn them upside down and plant into them!). While we talk about making things match, don't go too far. Too 'matchy-matchy' is just boring. Just what too far is will be different for every gardener. Ultimately, your garden is your happy place so go with what works for you, not what anyone else thinks works.

Paint

Paint can be very effective in creating a theme or breathing new life into old items. A chair, a gate, a feature pot and the letterbox all painted the same bright colour can be very dramatic, particularly in a garden without a lot of other colour.

In my front garden I have a red gate, a red letterbox, an old red fire hydrant and a red chair with a pot of succulents on it. Not only does this create a feature, it sets a theme. Many items can last much longer with a coat of paint to give protection from the elements, and this alone can make paint worth including in your garden creativity. Paint can be expensive, and it can be highly toxic. The eco-friendly paints tend to be even more expensive. There is an enormous amount of half used paint in tins in garages around the place. A few years ago I was working on a handful of projects and wanted to paint things. I put a post on social media asking if anyone had old paint. I got heaps! I was thrilled with the colours I ended up with as they suited what I was doing.

If the colours are not so exciting, can some extra tinting turn it into something brighter? Most leftover house paint will be neutral or pastel colours. Pale blue can be tinted to bright blue or purple. Pale yellow can be tinted to bright yellow or orange. White can be tinted to anything. I have mixed half tins of blue and yellow to make green. Obviously mixing tins of paint will not work if they are on different bases. Water based paints are much easier to use and more common, but oil based paints can be more durable for outdoor items in the garden. Using up old half tins of paint can save you a lot of money while keeping a problematic product out of landfill.

When painting your items for use in the garden, don't forget to paint the underside. Places like the bottom of chair legs or under seats get missed because we can't see them and without the protection of paint can get wet and rot.

Old tools

Don't throw away those old tools in the back of the shed. Hang them on the fence for effect or weld them together to make interesting sculptures, or onto a plain gate to make it more fun. Old rusty garden tools are really the perfect adornment for almost any garden.

Children's gardens

Children will be particularly keen to help with repurposing to create a garden they find exciting. Old toy trucks make good planters, perhaps with a toy dinosaur or two for added fun. Old crystal chandeliers that are falling apart are still very exciting for a fairy garden. A collection of buttons can be sprinkled over a gravel path for little fairies and little pirates to discover. This way those buttons get played with over and over. Eventually they will get lost in the gravel, but that's ok too. Let the kids' imaginations run wild! The garden doesn't have to be like that forever, but they will have had a great time in the meantime, and that is invaluable. I still randomly find old dinosaur toys around my garden, and it is a lovely reminder of when my son was little and gardening with me.

The kids may choose some items to use in the garden that are highly unappealing to you. Don't be too quick to say no. They won't be this age for long, and these unappealing items are not permanent additions to the garden.

Tyres

Tyres are popular in making raised gardens because they are easy to get for free and work very well. Filled with good soil and compost they can give both good drainage and pockets of retained water in the rims, which can be very effective. They are currently also highly controversial. Yes tyres do leach toxic compounds, so they are best used only for ornamentals and avoided for growing edibles.

Tyres leach most when they are shredded for recycling. By using the tyre intact, there are less leachates to worry about. Old tyres in the garden leach very small amounts over long periods of time. The best way to deal with this small load of toxicity is by adding compost – yes, if in doubt add compost! Soil microbes are important for breaking down many of these compounds in the soil. Tyres are absolutely no good for organic food growing, and not recommended even for non-organic food growing, but they can certainly have benefits when growing ornamentals.

They can be very useful to create mini garden beds around the base of trees in the lawn or on the nature strip. The tyre will protect the tree trunk from whipper snippers. If you are doing this though, make sure the tyre is larger than the expected mature size of the tree trunk. You don't want the tyre to end up strangling the base of the tree.

I do have tyres in my garden, all added over 15 years ago. I have used them to create raised beds for trees. These are only one tyre deep but in my heavy clay soil this allowed me to add enough compost and good soil to get the tree started, while also giving some protection from chooks and dogs. In a couple of places I have them two tyres deep to give drainage to grow succulents and other plants that hate wet feet. As I am not going for an auto theme, I have hidden them under trailing plants and by growing other plants in front of them. I don't really need them anymore but as I can't see them, there is no need to try and pull them out.

I have seen recommendation that you paint tyres to seal them and prevent leaching. This can be very useful in making them more visually appealing, but given that paint is also full of highly toxic compounds that also leach into the environment, it is hardly a solution. For many of us living in urban areas, a completely chemical free lifestyle is simple impossible. If a few old tyres in the garden allows you to grow plants where you otherwise couldn't this is an overall positive outcome. Keep up the composting so any toxins are dealt with and keep growing plants! The garden is still highly valuable even if we are not eating some parts of it.

Trellises

Old gates, ornate bed heads, even a series of bicycle wheels attached vertically to a pole can make artistic and interesting trellises. Part of the trick to making things like this work is weighing up the aesthetic and the functional value of an item. I have seen old springs from inner spring mattresses used beautifully as garden trellises, but equally I have seen them used where it looked like beans growing in a rubbish heap. As much as I am a huge fan of repurposing, it is best done with a little thought and attention to the aesthetics.

Features

A large urn with a chip in it can be turned so the chip doesn't show. A bird bath that no longer holds water can instead hold succulents or bromeliads. An ornate bird cage without a base can provide protection for plants from birds, possums or other unwelcome visitors. A garden gnome with a hat broken off can have a plant put in the top. An old cupboard can be converted to a characterful potting table, complete with a place to store pots and garden items. All of these once discarded items can have a new life as not just something extra in the garden but a feature item in your garden.

In my garden I have a feature created out of an old door with a tall candelabra holding up a chandelier given by a friend. Together they turn a tricky corner into a beautiful garden room. In another garden I have helped a client to create a garden around an old sewing machine. It was the first sewing machine she ever bought, with her first ever pay packet. It is now too old to be used, and while it is not an antique, it is vintage and holds a lot of precious memories. We have planted around it with colourful low growing plants to create a patchwork garden. Sure, it won't last forever but while it is there it will give enormous joy. It was otherwise taking up space in the back of the cupboard.

The greatest joy in repurposing is that you get to have so much fun creating your own unique items for the garden. Unleash your imagination and see what happens. If most of the items you use were discarded to begin with, it is no big deal to discard them again if they don't work for you, which means you have nothing to lose. They won't all last forever, but most of them won't need to. In the meantime, these items can be both fun to create and an ongoing joy to look at in the garden every day.

Using the Garden to reduce the Eco Footprint of Your Home

Modern living has seen a huge separation between the built and the natural environments. We are so busy controlling our artificially built environment that we forget to look to nature for answers.

I saw a huge house being built next door to a client's home recently. The house was so large it overshadowed the garden next door. The building process was not just noisy but involved earthshaking drilling, the removal of everything green and dense clouds of concrete dust. As it neared completion, gardens were planted on the very thin strips of bare earth that surrounded this enormous house. It wasn't until the air-conditioning bank was installed that we discovered that this house did not contain a single window that could be opened. This is an extreme, but very real and, I suspect, not isolated case. How this meets fire regulations I do not know.

Gardens get planted into tiny spaces now to make way for the huge houses people want. It seems to me that gardens are becoming something to look out on to, or are there as a privacy screen. In working with clients these days, so often the consultation starts indoors looking at the view from the windows, then going out and designing a garden to fit that view. I realise this is important when blocks are now so tiny there really isn't room for a garden, but it makes me sad that our appreciation for the garden is reduced to what is framed by the view from a window.

A client wanted a herb garden outside a long narrow kitchen window at bench height. From the kitchen you looked out onto herbs instead of the fence, which was great. Sadly this window did not open. It was not possible to reach out to pick or water the herbs, or even just smell them. The journey around from the kitchen to this herb garden was such that it did not get much care and the herbs did not get picked. We still want a connection to the world outside our home, but we have become very distanced from it.

This is not just an issue of time in nature (including the garden) being good for our mental and physical health, it is about the role of gardens to make our homes more liveable and energy efficient.

The idea that opening a window allows a cooling breeze in is pretty simple, and one most of us employ without a second thought. At least, it used to be. Orienting a house to make the most of these breezes is also sensible. Less understood is that if that breeze passes through or over lush green vegetation it is further cooled before entering our window.

Planting small trees, shrubs and other living plants around the house can help to cool the air around the house and the breezes coming in. Even air passing over green lawn will provide some cooling, especially if we compare that to air passing over a concrete driveway or main road before coming through our window.

Lawns can actually be rather beneficial in a modern garden. Too often the alternative to lawn is concrete, paving or gravel, not a green garden alternative. These hard surfaces are heat sinks and will store and reflect heat back onto your home. This could be highly desirable if you are trying to warm your home, but much less so if you wish to cool it. Even if you are trying to heat your home, you do need to be wary of large hardscaped surfaces. They adsorb and hold heat which is useful, but they also require a lot of heat to warm them up so can remain colder for longer when there is no source of heat (e.g., sunshine). Natural living surfaces are significantly better heat regulators for reducing the extremes of both heat and cold.

A well-placed shade tree will of course make a huge difference to the temperature of your home. You can make an even bigger difference by surrounding your home with natural green spaces, including lawn, instead of hard heat absorbing ones. A lawn does not need to be well maintained to benefit the environment, it just needs to be a living green space. (The earlier chapter 'Sustainable Lawn Care' is a useful guide.) In addition to cooling the air around your home, the lawn itself can be a lovely outdoor living space. Having a comfortable outdoor space to spend time in when the house heats up on a hot day will not just save on energy bills, it will be good for your mental health.

Just how much a lawn impacts on the cooling of your home is up for debate as most studies are done by lawn suppliers who have a vested interest. A study done in Israel suggested that an unshaded lawn had very little cooling impact. Even if having a living lawn does not actively cool the air in your environment, at the very least it won't increase temperatures in the way hard surfaces will.

Another way of using plants to modify temperatures is to use deciduous trees and vines. These can be positioned to provide welcome shade in summer but let the sun through in winter. A grape vine over a pergola can do this (in addition to providing fruit). A frangipani over a paved sitting area will also work. This is a beautiful way to regulate not just temperatures but also the amount of light coming through the windows at different times of the year.

Positioning trees or shrubs to shade western walls can make an enormous difference to the temperature inside the house. The hot sun hitting a wall, a window, or roof is a major contributing factor to the temperature inside the house, and something we can adjust with thoughtful planting. Obviously if you are going to plant a tree, you need to think about your choice of tree and where it goes. A bit of planning can go a long way when using the garden this way. Are you in a cold climate where warming is more important than cooling? Would a deciduous tree give the cooling you need in summer without making things too chilly in winter? Perhaps some tall shrubs would be enough, especially if they got a large annual haircut at the start of winter. The climate here in Brisbane is hot. Our summers are stinkers and winters are mild and short, so I encourage gardeners to think about staying cool in summer before worrying about what winter brings.

A local rental property recently cut down a glorious small paperbark tree on the front western side of the home. The house went from perfectly comfortable year-round, to the tenants desperate for air-conditioning, even with blinds on the windows. They moved out. It's not an uncommon story that as trees get removed air-conditioners get added.

Temperature regulation is the first thing that comes to mind when evaluating the role of green space in the eco footprint of a building, but it certainly is not the only benefit to having green space.

Living green plants with roots in the soil, be that trees, gardens or lawns, all provide protection from excess water runoff during rain events. This not only provides erosion control but reduces pressure on our stormwater systems, decreases flooding and through this, lessens the impact of drought. Yes – reduced drought! No, we have no magic formula to make it rain, and natural ecosystems are also not immune to drought, however the concrete jungle so many of us call home does exacerbate weather extremes.

The impact on drought is twofold. Firstly, with no green spaces, especially but not entirely limited to trees, there is greatly reduced transpiration. This means less water in the atmosphere and less chance of rain. It also means that when it does rain, chances are, the rain will follow the green corridors - like attracts like, especially water and the rain will follow where the air holds more water, which is above trees. Less rain will fall in the barren areas. When it does fall, the rain runs off the hard surfaces and into the stormwater system. The more of the natural rainfall that is diverted off the soil and into a system of pipes and drains that lead to a local waterway, the increased risk of both localised drought and flooding. Rainfall is naturally spread over an area and soaks into the ground over a large soil area. Restricting where this water can go will contribute to backups and flooding. With significantly less water being able to soak into the ground, subsoils become drier, and less able to sustain life.

Drought is officially declared when soil contains no plant available moisture to a depth of 60cm. This seems pretty extreme but think about our dense urban areas and cities – how does the water ever get that deep if it is always channelled into drains and sent out to sea? While I have discussed the role of hardscaping and water penetration in the landscaping section, here we see that the role of green space has compound impacts on climate – not just in our own homes and gardens but in our entire suburbs and beyond.

And we have not covered it all yet. Air pollution is a huge issue. During the 1960s in China lawns were seen as symbols of capitalism and many were removed. The upshot of this was such significant increases in dust, that they subsequently reversed this policy. Lawns and all green spaces act to suppress dust by protecting the soil surface, but also play a role in trapping air pollutants, dust included. Many indoor plants are now being promoted as indoor air purifiers. If one potted indoor plant can noticeably absorb indoor air pollutants, what effect will a full garden of plants have? I don't have a quantitative answer for you there, but you can bet your bottom dollar it's positive! This may not directly affect the eco footprint of our homes but it will directly affect our health, and the liveability of our homes.

All these advantages to having lawn area are magnified if you add trees to the picture. Trees play a huge role in cooling through shading, but they also play a very significant role in buffering wind and wild weather – that is, those trees that are well placed and correctly chosen so as not to fall on the house!

Trees in the garden

Trees often come up in conversations about gardening. They do a lot of good in a garden and, as I am sure you have figured out by now, I am a huge fan. But they are also the element within garden design that is likely to have the biggest impact. If you make a mistake with a tree, correcting it later can be expensive.

When choosing trees for a suburban garden the type of tree you choose will be critical. Take your time to make a decision as trees are a long-term planting and worth getting right from the start. A tree that grows to around five metres will suit most suburban gardens. Trees that reach eight to 10 metres will give more shade but need to be chosen and placed carefully. Aim to maximise the shade on living areas and hard surfaces, and in particular, think about shielding the hot afternoon sun. This usually means a north western aspect in the Southern Hemisphere.

For a tree this size, care also needs to be taken to ensure it is not subject to extreme winds and therefore not at risk of dropping branches or otherwise posing a safety threat. Good tree choice will make a difference – avoid soft or brittle species like eucalypts and poincianas.

Another strategy to maximise tree safety, particularly in relation to wind exposure, is to plant trees in clusters. Even a grouping of three small trees will reduce the exposure to wind (and therefore risk) that each tree individually receives in addition to enhancing the cooling effect of the trees. This is a more natural style of gardening than is planting in straight lines or having a single feature tree. It allows the trees to provide some protection to each other and to other plants growing beneath them. Your garden may not be especially wind prone, so you may think that wind is not a factor for you to consider. Most Australian gardens will have some storm risk, and with storm risk comes wind damage.

One of my much-loved small trees is a dombeya. In my garden it was getting broken each year as it came into flower at the same time that the westerlies arrived. I planted a pink Java cassia on the windward side and have had almost no damage to the dombeya since then. The pink Java cassia is a fast growing and strong tree, far better able to break up the force of the wind in the garden.

In many other ways the pink Java cassia was a poor tree choice for the space, and I did know that when I planted it, but I made an emotional rather than rational choice. It is now the most perfect tree. The size and shape shelter my front verandah and provide privacy from the neighbour's windows. It is deciduous so lets in winter light. It is a shame it is only half its mature size! This tree is going to need pruning at some point, and once you start pruning a tree, it becomes a regular job. A tree that has been lopped will never be as strong as one that has not been. The regrowth will need to be managed or you will increase your risk of broken branches.

This tree is now large enough to drop leaves in my gutters and by next year will be shading my solar panels. As much as I love this tree, I do need to think about its future. The larger it becomes, the more it will cost me to have it removed. It is time to think about a more suitable alternative and replace this tree before it becomes too big a job.

As much as I love trees, I do need to think sensibly about using them in the garden. If we all made careful tree choices, we would be able to fit more trees into urban gardens. The flow on benefits can be huge. Urban planners are starting to realise that trees in private gardens are a critical element to managing the urban heat island effect. Neighbourhoods with more trees are cooler and the people there are happier. They also tend to fetch significantly higher house prices.

When choosing the right tree for your garden there are a number of questions to ask about the tree, including:

- What is the mature height?
- What is the mature spread? (this indicates its overall shape)
- Do I need a deciduous tree?
- Does it have an aggressive root system?
- Will it suit my climate?
- Does it have a dense green canopy?

Trees have a very bad rap when it comes to roots and root damage. This again relates to making the wrong tree choice. A rule of thumb is that the larger the tree, the larger the root system. In most cases this will be enough to guide you in terms of root damage. A very large tree should be placed further away from pathways, buildings, retaining walls and other structures as its large root system will cause damage. A smaller tree, with a smaller root system will be less problematic. In most cases you can expect the root system of your tree to extend as far out from the trunk as does the tree canopy. We call this zone the drip line. This is the edge of the tree canopy and is where the rainfall tends to run off and water is most available, hence the roots like to be here.

There are always exceptions. Some trees such as liquid ambers and crepe myrtles will sucker, especially if they are regularly pruned. These suckers can extend well beyond the drip line of the tree.

Fig trees of almost all descriptions (except the edible fig, which is relatively well behaved) will send their roots in search of water with complete disregard for things such as drip lines or concrete. They cause enormous amounts of structural damage to pipes, driveways, retaining walls and even buildings. A homeowner nearby to me found the roots of a fig tree in the walls of her kitchen. The tree itself was nearly 20 meters away in another yard. This nightmare is easily avoided by choosing a more suitable tree.

I mentioned spread and shape in that list. If you wish to grow a tree with a lovely spreading canopy to give an umbrella like shade effect, think about where it will fit. If your space is up against your neighbour's fence this could be problematic. Trees overhanging fences tend to cause a lot of neighbourhood disputes. If you have a tight spot for your tree, chose one with a more upright shape so it is less likely to overhang your fence and become a problem for someone else.

I also mentioned the density of the canopy on that list. If the tree is for cooling value, choose one with dark green leaves and a dense canopy. They will offer deeper and cooler shade than will one with a sparser canopy and greyer leaves.

There are so many great trees to choose from. Once you work out the right sized tree for your garden, you can then think about things like what colour flowers would you like it to have? Do you want fruit?

In my small garden (600sqm) I have managed to fit in a golden penda (yellow flowers), powder puff lilly pilly (red flowers), two other lilly pillies, native daphne (white flowers), snow in summer (cream flowers), little Euodia (pink flowers), white oak (cream flowers), custard apple, flame tree (red flowers), large leaved and small leaves native tamarinds, pink Java cassia (pink flowers), several frangipanis and something else I don't know the name of. While this is more than I would recommend for most gardens, it does give me a delightfully sheltered and shaded space, while still receiving enough sun to have an abundance of flowers throughout the entire garden.

Once you have added a tree to your garden, you do need to care for it. Consider this. In nature, trees offer shelter, homes and food for wild animals big and small. In attracting animals into the immediate vicinity of the tree, those animals poo there, providing manure to help feed the tree. As the animal feeds on the tree, or shelters under it, so it returns the favour by way of manure. In our gardens this does continue to happen although on a smaller scale as we have reduced access to the tree by animals. In a garden we tend to think of trees as being completely self-sufficient. Few gardeners feed or water their trees. And yet in nature, manure and even water tends to accumulate around a tree. I have saved many a very sad tree just by advising gardeners to feed and water their tree. Where we no longer have kangaroos sleeping under the tree and bettongs fossicking around the base, we are no longer allowing nature to keep her tree well fed. We must step in to help. Grow a garden around your tree. As you care for the garden, you are also caring for the tree.

Sarah's Garden – More than just a garden, it's a way of life

Sarah's garden is not just hers alone. Four friends purchased a set of four units, with a shared garden. So often in shared garden spaces there becomes less ownership and the garden gets less attention than a standard private garden. That is very much not the case in Sarah's garden. This is a garden brimming with love, life and creativity. Sarah may be the lead designer and the driving force behind the garden but the other unit owners (and Sarah's partner Brian) are all gardeners as well so there is plenty of enthusiasm for this garden. Semi-communal living like this is not for everyone. I have personally never been very successful at sharing my garden with another who also wanted a say in what happened in the garden. But this group of friends have embraced negotiation to create a space which hugely enhances the liveability of their small units.

This glorious small garden has managed to fit in a tropical garden, a chook pen, an edible garden, outdoor clothesline, a pool pond, a trampoline on a lawn with sitting area and an outdoor dining area. Two of the units share a back deck that overlooks the beautiful garden. The people in the front two units do not have a deck or outdoor space so the back garden had to be designed such that different owners could use the garden at the same time without feeling like they were in each other's space.

Large tropical heliconias and bamboo have been used to screen one back corner from the house. A dining table and chairs were added to this hidden corner and it became a very well used section of the garden. A few 12V lights have meant the garden can be used at night and is the setting for many a communal dinner or party. This back corner is not only out of sight, it is out of reach of the greywater hose, so rarely if ever gets watered. Sarah has solved this problem by splashing out on gorgeous rusty steel edging to create a flowing tiered garden which has been planted with succulents. Sarah described this as her 'folly'. While this feature was not cheap, it looks sensational and will last a lifetime.

There is otherwise very little in the way of new materials used in the garden. Lots of second hand and found items have been used as garden edging, timber seats, and old pallets as timber cladding for the wicking gardens. Even the back fence is a 'found feature'. A collection of old art deco steel gates was hung as fence panels and painted bright blue. The plan here was to ensure the fence kept their chooks in and other people's animals out but did not block the view of the large park behind the garden.

The garden had to be able to be used as an extension of everyone's living space, and therefore it needed to have useable surfaces and shade for summer. Areas of paving were tried and removed as they stored too much heat. While the lawn requires mowing, a job no one really wants, it is a much cooler surface and therefore a far more enjoyable space for everyone to use. The lawn is never watered and is allowed to brown off in dry times.

Shade is provided by placement of clumping bamboo to block the western sun, and from a collection of native and fruit trees, like the old mulberry in another corner.

The chook pen is tucked in behind trees and created as the perfect place to grow vines. The vines are flowering and ornamental but also give shade and shelter for the chooks, which are naturally forest understory birds. The chook pen is in full view but the great shape, glorious old gate and the flowering vines make it a delightful visual feature. It is also the compost machine of the garden. Everything that fits gets put into the chook pen for the chooks to eat and otherwise break down. By composting in the chook pen, not only do the scraps break down quickly, it is enriched with manure making it the perfect soil enrichment for the garden. Sarah does occasionally use organic fertiliser pellets in the garden but the compost from the chook pen is the main source of goodness for her plants. The chook pen being in such a central location in the garden helps the girls be social and friendly and allows them to help with other garden tasks, like pest control. Sarah's daughter spends her mornings catching grasshoppers to feed the girls and makes weed salads for them.

Growing edibles proved to be a challenge, as with so many subtropical gardeners, the friends found themselves following planting advice from a different climate. Sarah openly states that her single biggest learning in the garden is to pay more attention to what condition each plant actually wants in order to grow well. Since she has made the effort to do this, she has had so much more success and is getting more joy (and more food) out of the garden. While they do not grow all of their own food, they find that by focusing on leafy greens and herbs they are able to provide what is needed fresh on a daily basis. Wicking gardens have helped to make growing edibles easier. They are a self-contained system so were initially put in to prevent root competition with the trees in the garden, but have since proven to be very efficient not just with water but also with nutrients. The first couple worked so well, that more went in and now everyone has their own wicking garden, giving everyone a little piece of garden to be just their own in a shared space.

The garden is watered almost entirely using rainwater and greywater. There are four 3000L rainwater tanks, and greywater systems connected to most of the showers and the laundry. The greywater has allowed them to grow very thirsty but much loved heliconias and bananas to their full potential. The garden has a very lush tropical feel, and the use of water pots adds to this. It is a cool and inviting place to be, even on a hot day.

Sarah has a passion for tropical destinations and for water features but had to express these passions in a small space and in a way that would be appreciated by others. There was no budget for a pool, and nor could they justify giving so much space to something that is used for only part of the year. First Sarah tried a couple of huge lotus pots. She got two so that she and Brian could each sit in one on a hot day with a beer. (I wish I had seen that!) Unfortunately, she found that when they sat in their pots, there wasn't enough room for fish, so they ended up with wrigglers instead, making them much less fun to sit in. So, the lotus pots grew lotus (from seed!) instead and the friends invested in a real pool, well, a real 16,000L above ground mini-pool which got sunk into the ground. Still, there was no way this space was going to be single purpose! Lots of plants were added to the pool, some planted on the bottom, but most in pots on the stairs.

In went fish, blue claw crabs and freshwater shrimp. And in went they. At times the water is so clear that they can even snorkel in there to watch the fish. At other times it is pretty murky, but still a great spot to jump in and cool off on a hot day, which they all do regularly.

They eat the blue claw crabs and some of the water plants such as kang kong and mint, but otherwise they just appreciate the habitat value. There is no water filtration system and no chemicals used so the pond is not crystal clear, but it is very alive and very much enjoyed by humans as well as wildlife. At different times they have noticed boom and bust cycles of different animals in the pond including dragonfly larvae, fish, shrimp, snails and tadpoles. It has given them the opportunity to watch an ecosystem and appreciate how they are both delicate and robust but most of all complex. The boom and bust cycles have never been treated as pest problems, and they have never resulted in reduced long term biodiversity in the pond. They are just part of what nature does as the weather changes.

This garden is very multi-purpose, not just in the amount of people it has to please but in all the functions it provides. It is beautiful, functional and comfortable. It provides food for humans and habitat for wildlife. It creates multiple outdoor living spaces which enormously increases the liveability of the four small units it surrounds. It even manages to combine a place for a refreshing dip with food production, beauty and a rich ecosystem. This is a garden which is a huge success on so many levels.

Becoming A Garden Ecologist

Emulating nature

If we want to learn how to garden with as little impact on the environment as possible, then who better to teach us than Mother Nature herself?

Ecology is the study of ecosystems – communities of plants and animals, and the interactions that make the entire system work. The inter-relationships between all species in an ecosystem are highly complex and we may never know the full extent of it, but the great thing is, we don't really need to. We just need to allow it to be.

By applying some ecological principles to our gardens, we find the garden can thrive with far less input from us. This happens when the various different plants and critters in the garden are working together, and allowed to do what comes naturally. Good bugs are there to keep the bad bugs in check. Plant eaters control excess growth and give our plants a prune, before they themselves are eaten by something else. Critters poo, turning the plant matter they have eaten into manure to fertilise the soil. Ants help to aerate the soil and remove weed seeds. Flies help to break down waste. This waste then enters the soil as organic matter, with the help of worms, and makes our garden grow better.

Monocultures are rare in nature. Where they do occur, it is usually in response to extreme conditions. Biodiversity is what nature strives for, a diversity in both the plant and the animal species which occupy an ecosystem. Everything has a place and a role in contributing to the value of the ecosystem.

In viewing our gardens as living ecosystems rather than just gardens, we start to see the interplay between the various plants, the critters and the inert landscaping elements, and we realise that our gardens are so much more than just a handful of plants positioned nicely for our own enjoyment.

Obviously gardens are not natural ecosystems – as in they would not occur naturally without our interference, but they are both valid and valuable ecosystems. As natural ecosystems are increasingly lost, unnatural ecosystems in the form of private gardens become increasingly critical in supporting wildlife and biodiversity.

Gardening schools such as organic gardening, permaculture and biodynamics have been espousing the use of natural processes for a long time now. I am a big advocate of using natural processes to do the work for you wherever possible. This is not to be interpreted as letting the weeds grow and the plants left unpruned and rampant – that is not gardening. We do play God in our gardens – we decide what to put where and what care to give. Natural process may do a lot of the work for us but not all. It is still up to us to put in the extra care which makes it a garden rather than a totally wild place. But by looking to those natural processes, we can have a wild place that looks like a garden.

Gardeners are emulating natural processes when we grow a tree or shrub to shelter the shade lovers beneath, or when we grow epiphytes on a trunk, practice companion planting, or plant special mixes to attract good bugs and grow comfrey for the compost value.

Some of these points may seem rather small and almost trivial – surely just doing a few big things like saving water is enough to call ourselves sustainable gardeners? What I am trying to show are some of the small details that are natural processes. They may be small jobs in the garden, but together they add up to a significant way of gardening, in which so much more can be achieved with much less effort or resource use. Once you get your head around the concept of the garden ecosystem, you will find your garden requires so much less maintenance, is healthier and happier, and is far more sustainable - all of which gives you so much more satisfaction as the gardener.

By choosing the best fit for you from the ideas presented in this book, you will find your garden evolves into a functioning and valuable little ecosystem. As this happens, there should be less work for you as the gardener and more time to observe and enjoy the garden. The reward is there not just for you, but for all of nature as well.

Notes

ABOUT THE AUTHOR

Kate Wall is a gardening professional based in Brisbane. Her career changed from environmental science (with scientific qualifications in Environmental Biology) to professional gardening after the 2011 floods when her local community was badly affected. Kate's volunteer efforts to restore flooded gardens saw her earn a number of awards including a prestigious 'Ray Phippard Fellowship' from the Lions International. This culture of caring has set the theme for Kate's approach to gardening as she works towards improving people's lives through making gardening a pleasure and a success.

Kate specialises in teaching people to garden in harmony with nature, ensuring beautiful and highly successful gardens for gardeners of all levels of experience. By working with nature Kate focuses on a very sustainable approach to gardening which is not only better environmentally but also makes the job at hand so much easier. She is a keen proponent of subtropical gardening. She encourages gardeners to understand and appreciate our unique climate and to use it to our advantage, regardless of the style of garden. Kate has her own highly successful subtropical cottage garden, producing food, herbal medicine and an incredible abundance of flowers and joy.

Kate is one of Australia's leading weed educators, and is the author of 'Working with Weeds'.
Kate is a past president of the Queensland Herb Society, and past vice president of the Horticultural Media Association of Queensland. She remains active in both organisations. Kate is also the creator and presenter of 'Gardening in the Pub'.

www.ingramcontent.com/pod-product-compliance
Lightning Source LLC
Chambersburg PA
CBHW060521010526
44107CB00060B/2647